On October 8, 1977, the Republic of South Africa issued the above stamp to celebrate 25 years of uranium development. It was 25 years before, on October 8, 1952, that the first uranium plant in South Africa opened at West Rand Consolidated Mines with a fixed bed strong base anion exchange resin technology.

Ion Exchange Resins in
Uranium Hydrometallurgy

Ion Exchange Resins in Uranium Hydrometallurgy

Ion Exchange Resins in Uranium Hydrometallurgy

E.J.Zaganiaris

consultant

Second Edition

IN THE SAME SERIES:

Ion Exchange Resins and Adsorbents in Chemical Processing, 2nd Edition, Books on Demand, 2016

Ion Exchange Resins and Synthetic Adsorbents in Food Processing, Books on Demand, 2011

The cover design was made by Dr María A. Pérez-Maciá

Published by:
Books on Demand GmbH,
12/14 rond point des Champs Elysées
75008 Paris, France

Printed by:
Books on Demand GmbH, Norderstedt, Germany

© Copyright 2017 : Emmanuel Zaganiaris
ISBN : 978-2-3221-5737-2
Dépôt légal: mai 2017

In loving memory of my wife

Preface

Since 1945 when the demand for uranium was high, it became clear that the available high grade uranium ores were no longer sufficient and that low grade ores had to be processed to cover the needs for uranium. This implied that large quantities of leach liquors had to be processed containing low concentrations of uranium together with other impurities. This made the classical method of direct precipitation unfeasible for economical and also for technical reasons. In the late 1940's strong base anion exchange resins were developed for recovering uranium from sulfuric acid leach liquors leading to the development of a process that became commercial in 1952 in South Africa. Fixed bed column techniques were used mainly, even though Resin-In-Pulp was also developed in the years 1950's, with baskets going up and down into the contactors, to reduce the cost of the solids-liquid separation. Continuous systems also started developing where liquors and resins moved counter-currently with the objective of decreasing liquid-solids separation costs, lowering resin inventories and reducing chemical consumption.

In the 1970's, continuous systems based on fluidized bed technologies developed in South Africa, led to their commercialization in many mines. In the 1980's and 1990's the uranium demand decreased considerably with the consequence that no new significant developments took place during these years. Since 2000 however, the decrease in uranium stocks and the increased need for nuclear fuel for power plants increased the need for uranium. This caused the price of uranium to surge to very high levels with the consequence that recently many projects for new mining sites or for new processing techniques in existing mines have arisen.

With the current expansion in uranium projects in perspective, this book summarizes past and current experience of using ion exchange technology in uranium recovery. The objective is to focus on ion exchange rather than on uranium hydrometallurgy and it is therefore addressed to people working in the uranium industry or in parallel industries such as engineering companies, equipment manufacturers and laboratories, for whom ion exchange is not their primary field of experience.

The first part of the book describes the structure and the properties of ion exchange resins, the principles upon which is based their performance and the processes in which this performance is obtained. This background is then applied to the second part of this book where it is specific to the recovery of uranium from acidic or alkaline leach liquors. Here, we discuss the chemistry involved, the main factors affecting resin performance, the equipment in which the resins are used, the main problems encountered and how to conduct laboratory evaluations of the resins used in these applications. Some simple examples of design calculations are also given.

This book results from my experience of working in the Ion Exchange department of Rohm and Haas Company and from the numerous visits to various uranium mines, engineering companies and institutes mainly in South Africa, Kazakhstan and Australia while in the employ of Rohm and Haas Company. Much information was obtained during these visits from discussions with metallurgists, engineers, scientists and technicians and this provided extremely valuable information.

Emmanuel J. Zaganiaris

Preface to the 2nd edition

Since the first edition of this book in 2009, membrane technologies start becoming part of the uranium processing but ion exchange continues to be the major player and new resins with better performance have been introduced. The first part of the book has been enriched, among others in the sections of equilibrium and selectivities of ion exchange resins. In the second part, more details and some new developments have been added like H_2SO_4 recovery from H_2SO_4 eluates using ion exchange, or the use of new products for high chloride content acid leach solutions.

EJZ

Acknowledgements

Acknowledgement is made to the NAC Kazatomprom, in particular Aleksandr P. Patrin of Geotechnoservice LLP, to Kemix (Pty) Ltd, in particular Mark Proudfoot and to Clean TeQ Ltd, in particular John Carr, for helpful information and permission to use diagrams included in the text.

Many thanks to Vassili Demidov of Rohm and Haas Russia (now Dow) for his valuable help in Kazakhstan.

I would like to express my thanks to Areski Rezkallah of Rohm and Haas, now in Lanxess, for his comments and discussions.

I am greatful to Peter Cable of PeterIan Consulting, for going through the draft of the first edition of this book and making many valuable and constructive comments.

Many thanks to Jaco Bester of The Dow Chemical Company for his helpful information and suggestions.

Finally many thanks are due to Dr María de los Ángeles Pérez Maciá of Dow Chemical for the design on the front cover as well as for numerous comments and suggestions.

ABBREVIATIONS

ADU	ammonium diuranate
AM	arithmetic mean
AMD	acid mine drainage
AMP	aminomethylphosphonic
BV	Bed Volume
BET	Brunauer, Emmett and Teller
DVB	divinylbenzene
CIX	Continuous ion exchange
CME	chloromethylether
DMA	dimethylamine
ES	Effective Size
GM	geometric mean
HM	harmonic mean
HMS	harmonic mean size
IDA	iminodiacetic
IEZ	ion exchange zone
IX	Ion exchange
IER	Ion exchange resin
ISL	in situ leaching
L_R	liters of resin
MHC	moisture holding capacity
MP	macroporous
MR	macroreticular
NIM	National Institute for Metallurgy
NF	nanofiltration
PLS	pregnant leach solution
RIL	Resin-In-Leach

RIP	Resin-in-Pulp
S	styrene
SAC	Strong acid cation
SBA	Strong base anion
SDU	sodium diuranate
SIR	solvent impregnated resins
SX	solvent extraction
TMA	trimethylamine
TWD	true wet density
UC	Uniformity Coefficient
Vol cap	volume capacity
WAC	Weak acid cation
WBA	Weak base anion
Wt cap	weight capacity

CONTENTS

Preface
Acknowlegments
Abbreviations

Part I. Principles of Ion Exchange

1. Introduction 19
2. Structure of IER 22
3. Properties 39
   - Moisture content 39
   - Total exchange capacity 44
   - Density 46
   - Particle size 48
   - Pressure drop...................................... 51
   - Resin fluidization 56
   - Porosity, pore size, pore size distribution and internal surface area 59
   - Physical and chemical stability 62
   - Dissociation of weak electrolytes 70
4. Equilibrium and kinetics of ion exchange 74
   - Ion exchange equilibrium 74
   - Selectivities 87
   - Kinetics 95
5. Ion exchange processes........................... 101

Batch (equilibrium) operation	101
Column operation	111
Single column fixed bed	111
Loading cycle	114
Regeneration cycle	125
Three column merry-go-round system	131
Continuous systems	134

Part II. Uranium recovery

6. Uranium extraction	137
Complex-ion equilibrium	146
7. Acid leaching	151
Chemistry of acid leach	151
Absorption of uranium	154
Effect of resin particle size	156
Effect of uranium concentration	159
Effect of pH	160
Effect of chlorides and nitrates	161
Effect of iron	163
Thorium	164
Molybdenum	164
Vanadium	165
Elution	165
Recovery of uranium from H_2SO_4 eluates	171
Removal of uranium from mining wastes	175
Acid Mine Drainage	175
Contaminated water from mining sites	177
8. Alkaline leaching	178
Chemistry of carbonate leach	178
Absorption of uranium	179

      Elution 181
      Recovery of uranium with carboxylic resins ... 182
        Removal of uranium from drinking water... 183
9. Unconventional uranium resources 185
      Uranium recovery from H_3PO_4 185
      Uranium from seawater 187
10. Ion exchange systems in uranium recovery 190
      Fixed bed three columns in series 190
      Continuous ion exchange 195
        Fluidized bed 196
        Moving packed bed 201
        Resin-in-Pulp 209
11. Resin fouling 218
      Silica 219
      Polythionates 221
      Cobalt 222
      Molybdenum 222
      Caustic regeneration 223
12. Laboratory resin evaluation 225
      Preliminary resin preparation 226
      Column tests 229
        Charging ion exchange resins into columns 229
        Exhaustion (loading) cycle 230
        Elution 234
      Batch experiments 235
        Equilibrium tests 235
        Rate of loading tests 237

References 240

Subject index 252

The uranium fission reaction was discovered in 1939 by Otto Hahn and Fritz Strassmann. The above stamp, depicting the fission reaction, was issued by the German Democratic Republic in 1979 on the occasion of the 100th anniversary of Otto Hahn's birth.

Part I. Principles of Ion Exchange

1. Introduction

Ion exchange resins (IER) and polymeric adsorbents is a class of products that constitute a part of the separation technologies, like distillation, adsorption on activated carbon, or reverse osmosis, ultra filtration and electrodialysis membranes.
Ion exchange resins have numerous industrial applications such as:
Water treatment: industrial, ultra pure and potable water;
Food industry : sugar and other sweeteners (glucose, fructose, polyols), milk whey, purification of organic acids such as citric, lactic, glutamic, malic and maleic acids, fruit juices, amino acids; Processing of pharmaceuticals : antibiotics, vitamins;
Waste water treatment : removal of heavy metals and organics (such as phenol, benzene, toluene, xylene and others);
Purification of chemicals : brine purification in the chloralkali industry, purification of hydrogen peroxide, glycerol and others ;

Hydrometallurgy : uranium, gold, platinum group metals and base metals recovery.
Other particular applications are in chromatographic separations and purifications, or as an organic catalyst in chemical reactions, where their function is not as ion exchangers but as a solid acid or base.

IER are crosslinked polymers containing ionic functional groups attached to the polymer matrix by covalent bonds. In order to maintain neutrality, an equal number of mobile ions of opposite charge, called counter ions, are also present. The most common ionic groups grafted on the polymer matrix found today commercially are sulfonic, carboxylic, quaternary ammonium, and tertiary ammonium, groups. The IER are then strong acid cation (SAC), weak acid cation (WAC), strong base anion (SBA) and weak base anion (WBA) exchangers, respectively. Similarly, the most common polymer matrices found commercially are styrenic, acrylic and phenolic.

Also classified under IER are polymers containing ligands as functional groups. Such ligands include iminodiacetic (IDA), aminomethylphosphonic (AMP), thiol, thiourea and other. Polymeric adsorbents consist of a polymer matrix as with the IER, but without functional groups, or containing a small quantity of functional groups in order to give to the polymer matrix certain properties, for example hydrophilicity. They have a very large surface area in comparison to the conventional IER which allows them to adsorb relatively large quantities of organic molecules.

In a nutshell, when a solution of an electrolyte comes into contact with an IER, ions of a given charge are fixed by the resin while an equivalent quantity of ions of the same charge are released by the resin into the solution. On this basis, one can use resins to remove all ions from a solution, thus purifying the solvent (water in most cases), to capture and recover some of the ions by eluting the resins, or to remove specific ions and eventually replacing them with other ions in order to give the desired composition to the solution. Other mechanisms of removing ions or molecules from a solution with IER include adsorption, exchange of ligands and formation of complexes.

The following is a brief description of the structure and the properties of IER, together with the basic principles of ion exchange. The objective is to assist the reader in understanding the second part of this book where the application of IER in uranium recovery is discussed. There are many excellent references, books and articles on ion exchange and some general references are given in the bibliography at the end of the book (Kunin, 1958; Helfferich, 1962; Arden, 1968; de Dardel and Arden, 1989; Clifford, 1999).

2. Structure of IER

In general, IER are spherical in form with a few products made in granular form. Figure 2.1 gives an illustration of a typical IER bead. It consists of a polymer matrix, a crosslinking agent, the functional groups and the counter ions.

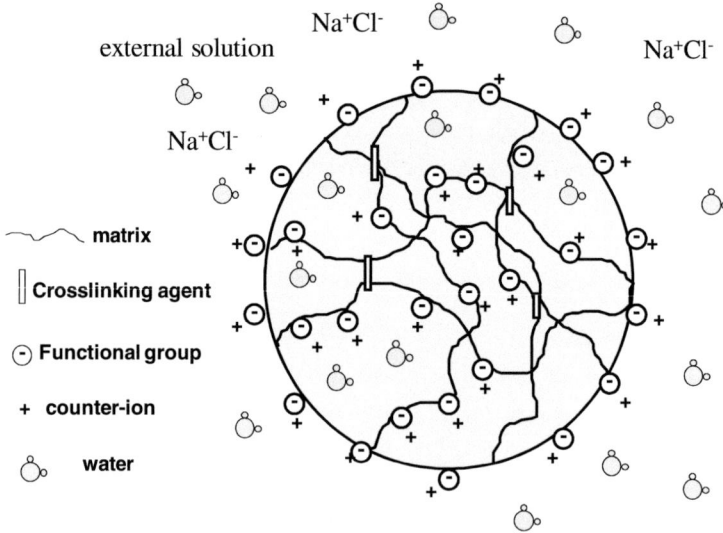

Figure 2.1 Schematic representation of an IER bead

In this case the functional groups are negatively charged, it could be for example sulfonic groups, $-SO_3^-$. The counter-ions could be H^+, Na^+, K^+, Ca^{2+} etc or combinations of them. The monomers that are used to form the polymer matrix of the IER are styrene (S), divinylbenzene (DVB) or acrylic esters. Polystyrenic resins are the most common types. These are co-polymers of styrene and divinylbenzene where divinylbenzene is the crosslinking agent. Another family of resins is made by polymerizing acrylic esters with divinylbenzene . A third type are the phenolic resins which are condensation polymers of phenol and formaldehyde (Fig. 2.2).

The polymerization of styrene with divinylbenzene and of acrylic esters with divinylbenzene is a free radical polymerization and is carried out in suspension in aqueous media. The mixture of the monomers is dispersed in water using dispersing agents and stabilizers to stabilize the monomer droplets and avoid collapsing, and an initiator is added to start the polymerization reaction. The polymerization is carried out under carefully controlled agitation and temperature conditions and over a certain time. The copolymer obtained in this way has a spherical shape.

When polymerization is carried out in conventional reactors under stirring (suspension polymerization), the polymer beads obtained have a wide particle size distribution. This can be controlled to a certain degree by variations and control of agitation and the type of suspension media. Alternate technologies exist which result in a narrow particle size distribution of beads. These uniform particle size resins are available under different trademarks such as Amberjet™ of Rohm and Haas Co. (now The Cow Chemical Company), Dowex® Monospheres® of The Dow Chemical Company or Lewatit® MonoPlus™ of Lanxess.

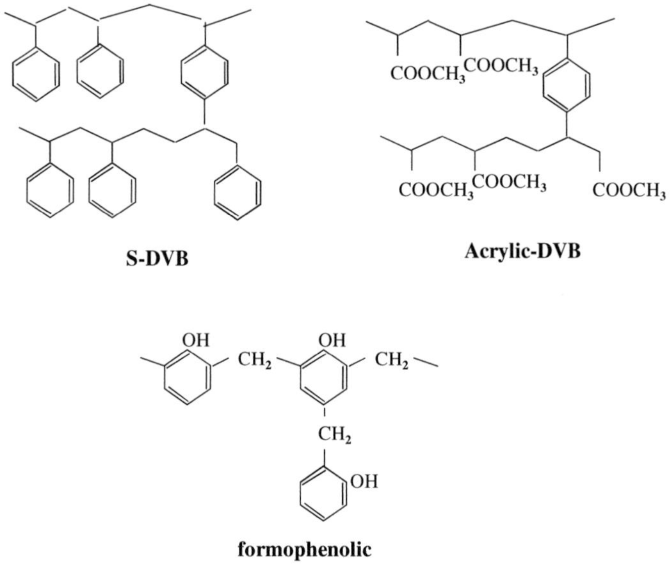

S-DVB

Acrylic-DVB

formophenolic

Figure 2.2 polymer matrices of IER

The copolymers formed are homogeneous and the final products derived from these copolymers are called gel-type resins. In another type of resins, the mixture of the monomers contains a porogen which creates macropores. In one of the earliest manufacturing processes (Bortnick, 1962; Corte and Meyer, 1972), this porogen is miscible with the monomers but is a non-solvent for the formed copolymer. When polymerization takes place, the formed polymer precipitates out and forms an agglomerate of microbeads glued to each other. The size of these microbeads depends on the polymerization conditions (composition of the mixture monomer- divinylbenzene - porogen, the initiator concentration, temperature). The porogen used during polymerization is removed either by distillation or by additional washing leaving a certain macroporosity in the resin particle. The copolymer beads thus formed have macropores varying from some tens of Angstroms to several hundreds of Angstroms in size. The resins made with this type of copolymers are called macroporous (MP) or macroreticular (MR) ion exchange resins. Depending on their size, the pores (pore diameter) are classified as micropores, mesopores and macropores, defined as follows:

        Micropores <20 Angströms
        Mesopores 20 – 500 Angströms
        Macropores > 500 Angströms

Following the formation of these various copolymers, which are solid, hydrophobic beads or granules, these beads or granules are washed to remove the residues of the dispersing media, monomers and other impurities. If porogen was used during polymerization, this is removed either by distillation or by additional washing. Then the beads are dried, screened and then, eventually, are activated by grafting the functional groups. Synthetic adsorbents are macroporous, or macroreticular, polymers bearing no functional groups.
Figure 2.3 schematically illustrates one bead of a macroporous type copolymer.

Phenol-formaldehyde resins are made by bulk polymerization in which the polymer is formed as a block after which it is broken up, ground and sieved to size. The copolymer particles formed with this technique are not spherical but granular in form. These copolymers even though they do not use a porogen during manufacture, also contain a certain form of macroporosity.

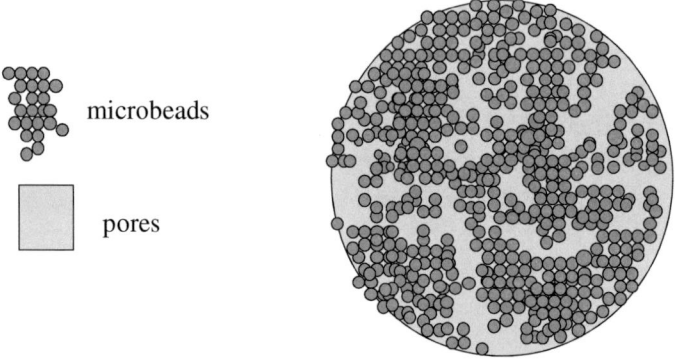

Figure 2.3: Schematic representation of an MR type resin bead

Today, certain adsorbents are made with a different technique. Davankov and co-workers (Davankov and Tsyurupa, 1989) developed a new type of adsorbents where the porosity was introduced not during polymerization but during a post-crosslinking step. To the mixture of the monomers a diluent is used that is a good solvent for the formed polymer. After polymerization, the polymer, either linear or highly swollen polymer is post-crosslinked with an external crosslinking agent where it is obtained a rigid hypercrosslinked structure having a very high surface area (800-1500 m^2/g) and a small average pore diameter, in the range of 10-40 Å. These adsorbents are called Hypersol-Macronet^{TM} sorbents (Purolite International Ltd.).

In comparison to the MR types described above, which have micro-, meso- and macropores in different proportions according to the manufacturing process, the hypercrosslinked structures would have many small size pores but at the same time large pores which facilitate the access of the small pores. They would look like in figure 2.4.

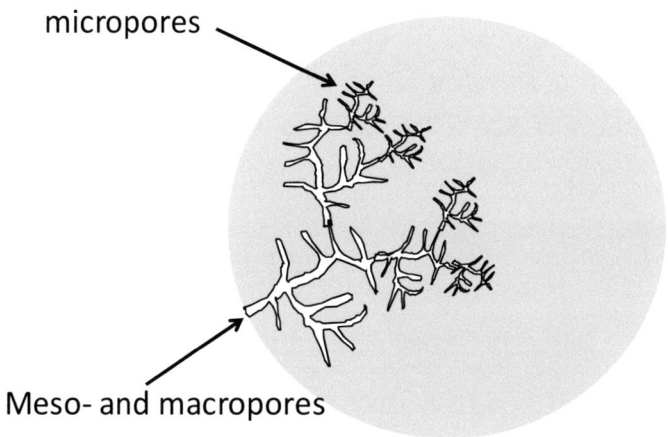

Figure 2.4 Schematic representation of hypercrosslinked polymers

The functional groups of the IER are ionic groups that give the copolymer beads the ability to exchange ions, hence the name ion exchange resins. Depending on the ionic group grafted on the polymer matrix, the resins are either strong acid cation exchangers, weak acid cation exchangers, strong base anion exchangers or weak base anion exchangers.

The SAC exchange resins typically have a sulfonic group as the functional group and are made by treating the different copolymers with concentrated sulfuric acid (fig. 2.5).

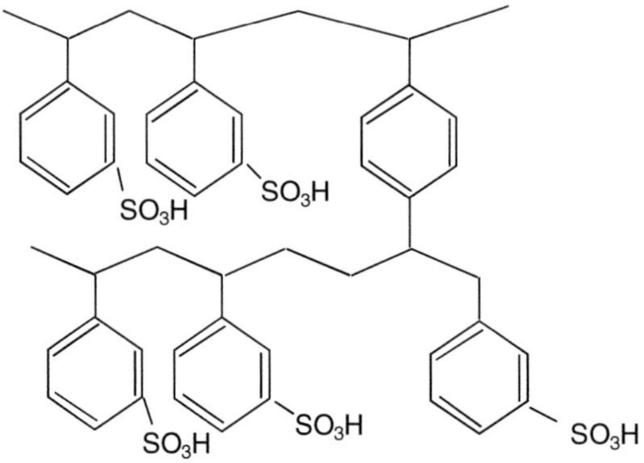

Fig. 2.5 Strong acid cation exchange resins

After sulfonation, the beads are carefully hydrated to avoid excessive exothermic reaction and then washed to remove unreacted sulfuric acid to condition the resin for use. The SAC resins made this way are supplied in the H^+ form. Different ionic forms such as Na^+ form, Ca^{2+} form etc are also available and thus these SAC resins are converted from the manufactured H^+ form to different ionic forms for appropriate use.

The weak acid cation exchangers are made from acrylic copolymers by hydrolyzing the ester groups. Thus, the functional group here is the carboxylic acid (-COOH) group which is a weak electrolyte. Contrary to the sulfonic resins which are strong electrolytes and which are able to exchange ions under almost any pH environment, carboxylic resins function only under certain conditions, as discussed later. If the hydrolysis of the acrylic esters is carried out under acidic conditions, the resin will be in the H^+ form. If it is carried out under alkaline conditions the resin will be in the Na^+ form.

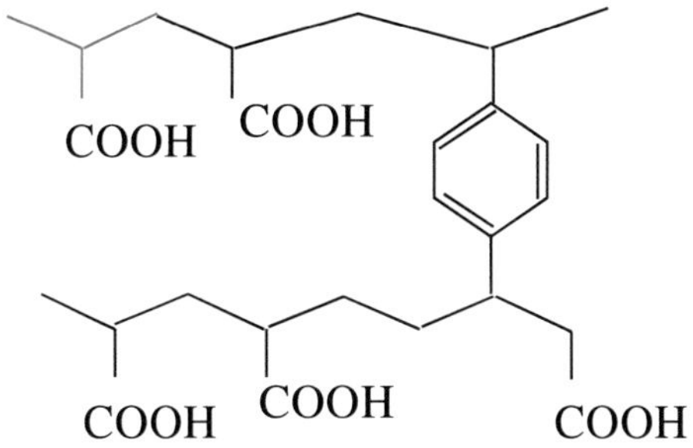

Fig. 2.6 Weak acid cation exchange resins

An alternate method for the manufacture of carboxylic resin is the hydrolysis of a polyacrylonitrile-divinylbenzene copolymer by washing to remove all $-NH_4^+$ ions formed during this hydrolysis and to convert the resin into the H^+ form. The final products made by either copolymer routes are the same (fig. 2.6)

Strong base anion exchange resins have quaternary ammonium groups grafted onto a styrene-divinylbenzene or on an acrylic-divinylbenzene copolymer. SBA resins exist in two forms, Type

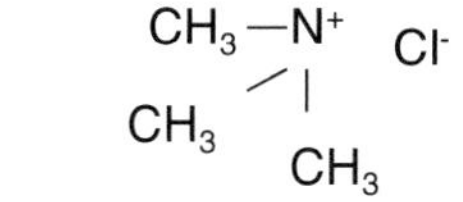

Type 1 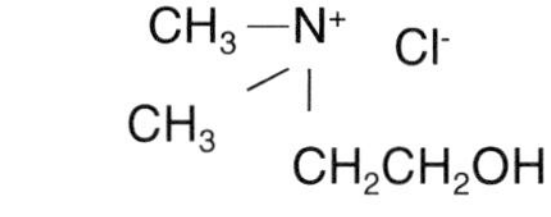   Type 2

Figure 2.7. Strong base anion exchange resins of Type 1 and Type 2

1 where the functional group is trimethyl benzyl ammonium and Type 2 where the functional group is dimethyl hydroxyethyl benzyl ammonium (fig. 2.7). There exist other types of SBA exchangers like triethyl, tributyl, dimethylpropanolamine (Type 3) (Dale and Irving, 1992), or bifunctional like triethyl/trihexylamine groups (Gu et al, 1999; Alexandratos et al, 2000), as well as with long aliphatic chains between the benzene ring and the quaternary amine group (fig. 3.12 page 67) with the purpose to amplify the affinity for monovalent anions over polyvalent ones or to improve their thermal stability.

The synthesis of styrene-divinylbenzene anion exchange resins is more complex than that for cation exchangers. First, the co-polymer is reacted with chloromethylether (CME) to chloromethylate the benzene rings (fig. 2.8).
The CME reagent as well as some by-products formed are highly toxic and dangerous for health and because of that, this reaction takes place in special installations where the atmosphere is completely controlled and monitored.

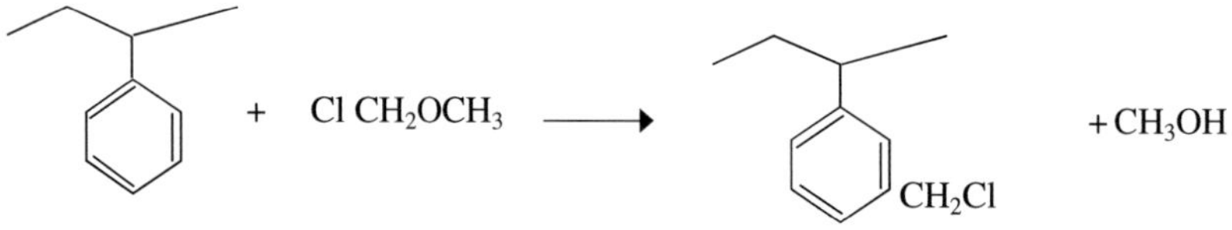

$$\text{[styrene]} + Cl\,CH_2OCH_3 \longrightarrow \text{[chloromethylated styrene]}\,CH_2Cl + CH_3OH$$

Figure 2.8 Chloromethylation

The chloromethylated beads are then washed to stop the reaction and to destroy any unreacted CME reagent. The intermediate chloromethylated beads are then available for amination. Amination consists in reacting with this intermediate chloromethylated bead with different amines: dimethylamine (DMA) to form weak base anion exchangers (fig. 2.9) or trimethylamine (TMA) to form Type 1 SBA exchangers or with dimethylhydroxyethylamine to form Type 2 SBA exchangers.

There are also two side reactions which can take place. One is the formation of methylene bridging either within the same polymer chain or between two polymer chains and these chains then become crosslinked (fig. 2.10). This is called secondary crosslinking to distinguish from the primary crosslinking through the divinylbenzene .

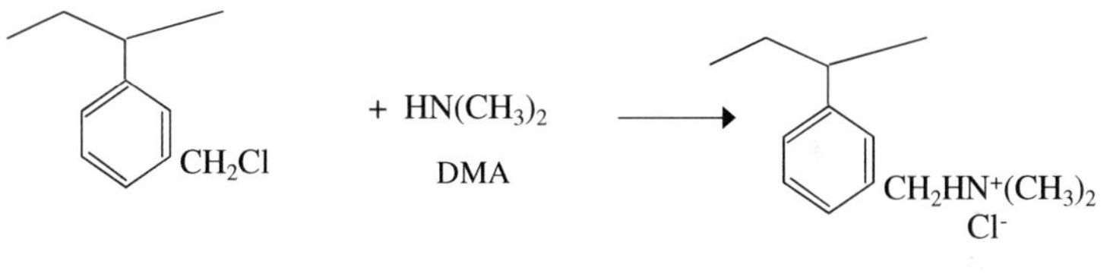

$$\text{Ar-CH}_2\text{Cl} + \text{HN(CH}_3)_2 \xrightarrow{\text{DMA}} \text{Ar-CH}_2\text{HN}^+(\text{CH}_3)_2\ \text{Cl}^-$$

Figure 2.9 Amination

If this methylene bridging reaction is uncontrolled, amination of the chloromethylated groups is affected and this can result in a final resin having somewhat lower capacity than the desired capacity.

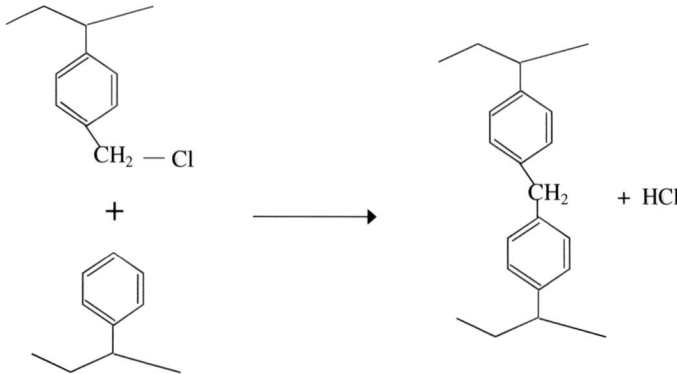

Figure 2.10 Secondary crosslinking by methylene bridging

In another side reaction, a weak base amine group can react with a chloromethylated group to form a strong base group and at the same time some additional secondary crosslinking (fig. 2.11). Because of this side reaction, the styrene-divinylbenzene WBA resins made with this chemistry contain a certain strong base capacity.

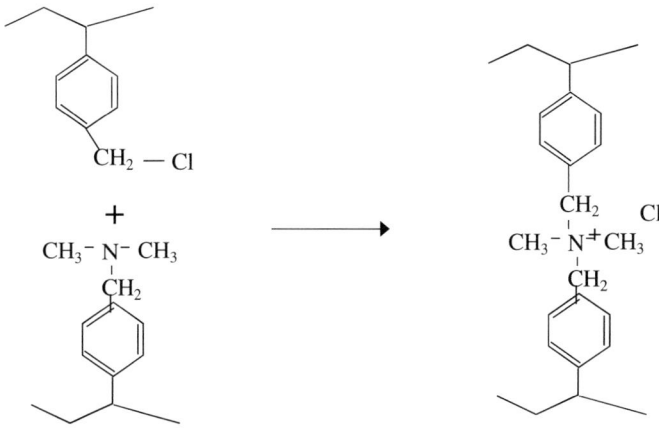

Figure 2.11 formation of strong base groups in WBA resins

When using acrylic-divinylbenzene copolymers, the WBA anion exchange group is formed by reacting the copolymer with dimethylaminepropylamine. In this way, the amine group is grafted to the polymer backbone via an amide group. The corresponding SBA exchanger is obtained from the acrylic-divinylbenzene WBA resin by reacting with it with methylchloride (fig. 2.12).

Because of the different chemistry, the acrylic WBA resins do not contain any strong base groups and are 100% WBA exchange resins. This is an important property as some applications cannot tolerate any strong base functional groups.

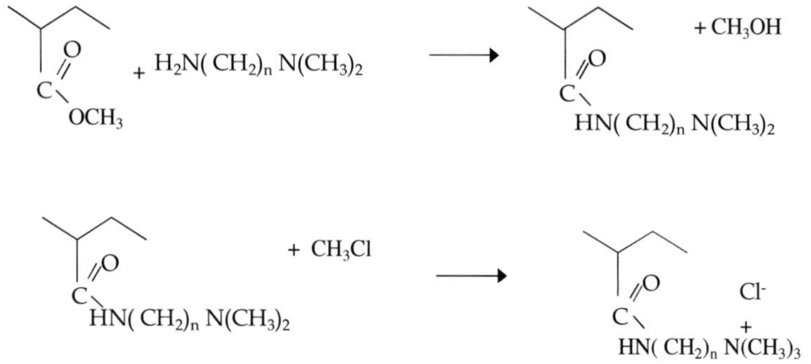

$$\text{R-COOCH}_3 + \text{H}_2\text{N(CH}_2)_n\text{N(CH}_3)_2 \longrightarrow \text{R-CO-NH(CH}_2)_n\text{N(CH}_3)_2 + \text{CH}_3\text{OH}$$

$$\text{R-CO-NH(CH}_2)_n\text{N(CH}_3)_2 + \text{CH}_3\text{Cl} \longrightarrow \text{R-CO-NH(CH}_2)_n\overset{+}{\text{N}}(\text{CH}_3)_3 \;\; \text{Cl}^-$$

Figure 2.12 Weak and strong base acrylic resins

The phenolic anion exchange resins found in the commerce are weak base and are made by reacting the copolymer with a polyamine such as triethylene tetramine (TETA) (fig. 2.13).

This polyamine nature gives these resins some unique properties such as the capability of forming complexes with transition metals that form ammine complexes such as Ni^{2+}, Co^{2+}, Cu^{2+} etc. With this mechanism, which strictly speaking is not ion exchange but a complex formation, these weak base resins can selectively remove these transition metals without changing the nature of the solution being treated.

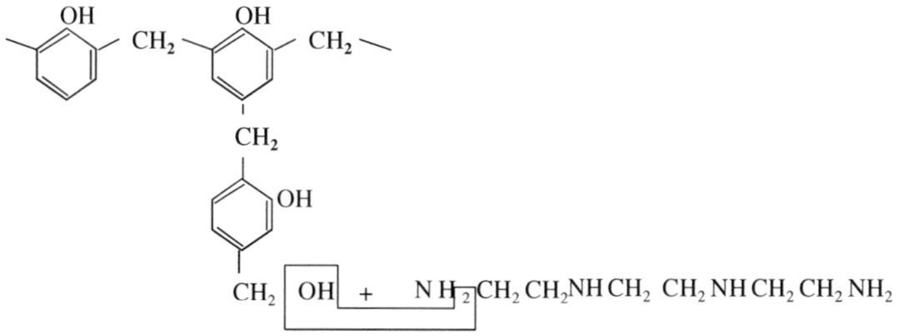

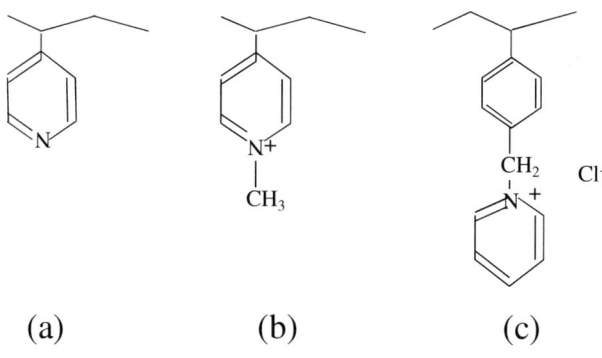

Figure 2.13 Formophenolic weak base resins

Another type of anion exchange resins is based on pyridine. It can be either a 4-vinylpyridine/DVB copolymer (fig. 2.14a) which subsequently can be quaternized to form a strong base anion exchanger (fig. 2.14b), or having a benzylpyridinium functional group (fig. 2.14c).

Figure 2.14 Anion exchangers based on pyridine

In addition to the above types of IER, there are also some "special resins" which have special functional groups, either grafted or attached to the polymer matrix and which function by forming complexes with metals in solution or by other special interactions. Because of this, these resins can be very selective for certain elements even in presence of high concentrations of other elements bearing the same charge. They include the following types:

Selective resins: They are styrene-divinylbenzene polymers bearing groups (ligands) capable of forming complexes with transition or other metals. If the ligands are bidentates or higher, these resins are called *chelating resins*. The most common functional groups that form strong complexes with uranium are:

Iminodiacetic (IDA) type functional groups, forming complexes with two oxygen and one nitrogen electron pairs:

$$\text{--[\text{Ar}]}_n\text{--CH}_2\text{--N:} \begin{array}{c} \text{CH}_2\text{COO:}^- \\ \searrow \\ \nearrow \\ \text{CH}_2\text{COO:}^- \end{array} \rightarrow M^{2+}$$

IDA resins form strong complexes with alkali earth and with transition metals. These resins work effectively at pH values above 2, the exact pH value depending on the metal concerned. The approximate selectivity sequence of these resins is as follows:

$Fe^{3+} > Cu^{2+} > H^+ > Hg^{2+} > UO_2^{2+} > Pb^{2+} > Ni^{2+} > Zn^{2+} > Cd^{2+} > Co^{2+} > Fe^{2+} > Mn^{2+} > Ca^{2+} > Mg^{2+} > Sr^{2+} > Ba^{2+} >>>> Na^+$

Aminomethylphosphonic (AMP) type functional groups:

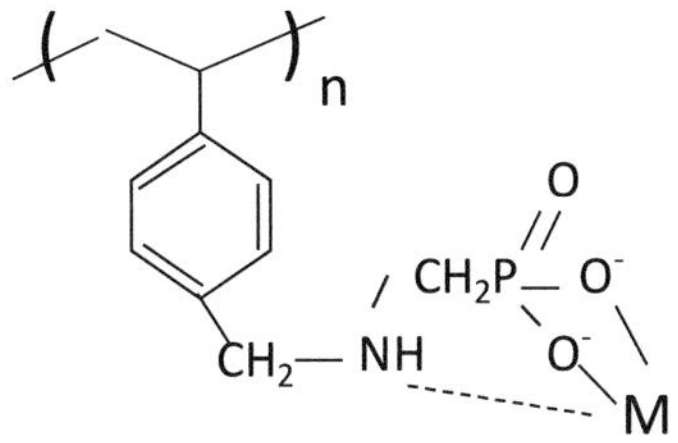

AMP resins form strong complexes with alkali earth elements and transition metals but also with uranium dioxide (UO_2^{2+}) and can be used to recover uranium from wet phosphoric acid, from acid leach pregnant leach solutions (PLS) containing high Cl^- levels, from H_2SO_4 eluates or from seawater.
The selectivity sequence of AMP resins for different metals is:

$UO_2^{2+} > Pb^{2+} > Cu^{2+} > Zn^{2+} > Ni^{2+} > Cd^{2+} > Co^{2+} > Ca^{2+} > Mg^{2+} > Sr^{2+} > Ba^{2+} > Na^+$

Amidoxime resins made from polyacrylonitrile/DVB copolymer by reacting with hydroxylamine (Hubicki and Kołodyńska, 2012). The complex of amidoxime groups with metal cations is shown below:

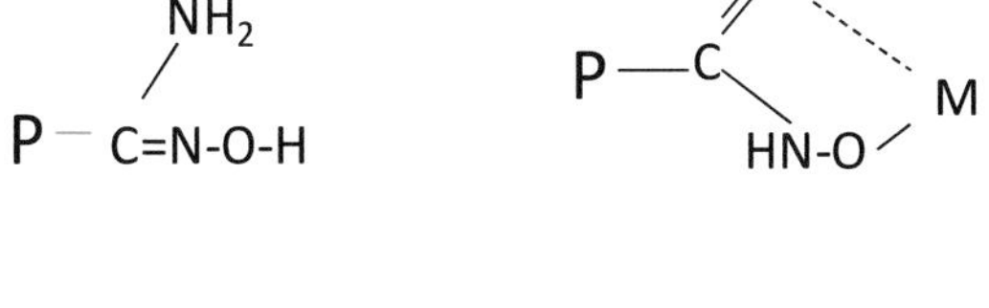

Amidoxime resins have been used to recover uranium from sea water and gallium from the bayer liquors. It shows also affinity for As^{3+}, Cu^{2+}, Pb^{2+}, Cd^{2+} and Fe^{3+}.
Other chelating resins have thiol, thiouronium, thiourea or bispicolylamine functional groups.

Among the special resins are *Diphonix® resins* (made by Eichrom Technologies Inc.) which are styrene-divinylbenzene copolymers having sulfonic and diphosphonic functional groups on the same matrix. The diphosphonic groups contribute to the selectivity of these resins for metals while the sulfonic groups give hydrophilicity to the resin and enhances the accessibility of the diphosphonic groups.

Another class of special resins developed by IBC Advance Technologies and promoted under the name of SuperLig^{TM} is used in *Molecular Recognition Technology (MRT)* Applications. These are very selective materials bearing specially designed ligands, such as macrocycles, bonded or not on solid supports such as silica gels or polymers.

Finally, another example of chelating products is liquid extractants adsorbed on polymeric adsorbents and are then used as solid ion exchangers. These products are known as *Solvent Impregnated Resins (SIR)* and contain such solvents as di(2-ethylhexyl) phosphoric acid (D2EHPA) among others. SIR containing various solvents have been used to recover metals from aqueous solutions and eventually separate them by selective elution using different eluents.

3. Properties of ion exchange resins

The properties of IER which are examined below have a direct bearing to the structure of the resins and directly or indirectly determine their function in the various ion exchange applications mentioned briefly in the introduction.

Swelling, Moisture content

If a dry or partially dry resin comes in contact with water, the water from outside the resin beads will enter into the resin and the resin will swell (Helfferich, 1962). This swelling results from the tendency of the hydrophilic fixed and mobile ionic groups to become hydrated making the polymer chains to expand. Also, since the ionic concentration inside the resin is higher than the ionic concentration in the external solution, water enters into the resin to balance the osmotic pressure difference between the interior of the resin and the external water. A third reason for the swelling of the resin is the electrostatic repulsion of fixed neighbouring ionic groups. The swollen resin will approach equilibrium with the outside water as the above stretching forces (hydration of ions, osmotic pressure difference and the increasing distance of neighbouring ionic groups) decrease. The water contained in a resin in equilibrium with pure water is called moisture holding capacity (MHC), or water uptake or other similar term, and is one of the principal characteristic properties of the resin. The moisture holding capacity (MHC) of the resin is defined as the quantity of water in grams that 100 g of drained wet resin under controlled atmospheric conditions (100% humidity in the air) will contain. This is expressed as %.

$$\% \text{ MHC} = \frac{[(\text{weight of wet resin}) - (\text{weight of dry resin})]}{(\text{weight of wet resin})} * 100$$

The solids fraction of a resin is then equal to:

$$\text{Solids fraction} = (100 - \text{MHC})/100 \qquad (3.1)$$

The MHC of gel type resins depends upon the crosslinking density of the polymer phase, the nature of the polymer phase (aromatic, acrylic), the ionic functional groups concentration (the total exchange capacity) and the nature of the counter-ions. The MHC of MR resins is the sum of the water found in the macropores and that found in the polymer phase. Thus, the MHC of MR resins is a combination of gel phase moisture and macroporosity moisture.

The effect of the nature of the counter-ions on the degree of swelling is somewhat complex. In general, the resin expands when one counter-ion is replaced by another one of bigger size in the hydrated form. For example, the size of the hydrated cations (table 2.1) varies in the following order:

$$H^+ > Li^+ > Na^+ > K^+ > Rb^+ = Cs^+$$

This is therefore the sequence of increasing resin volume when the resin is converted to the corresponding form.

TABLE 3.1 Ionic radii of hydrated ions

Ion	Ionic radius (A)	Ion	Ionic radius (A)
H_3O^+	2.80	Be^{2+}	4.59
Li^+	3.82	Mg^{2+}	4.28
Na^+	3.58	Ca^{2+}	4.12
K^+	3.31	F^-	3.52
Cs^+	3.29	Cl^-	3.32
Ag^+	3.41	Br^-	3.30
NH_4^+	3.31	I^-	3.31

From Volkov and Deamer, Eds, Liquid-liquid interfaces, CRC Press, Boca Raton, FL, 1996

The hydrated diameters for Mg^{2+}, Ca^{2+} and Sr^{2+} are 8, 6 and 5 A respectively (from Kielland, 1937) so that in the above Table 2.1, Sr^{2+} should be situated after Ca^{2+}.

A sulfonic SAC resin therefore swells going from the Na^+ form to the H^+ form and shrinks going from the Na^+ to the K^+ form. Inversely, it shrinks going from the H^+ form to any other cation form in this list. Similarly, SBA resins shrink going from the

OH⁻ form to any other anionic form. Some swelling figures for SAC and SBA resins are given below:

SAC resins	8% DVB	10% DVB
Ion	Swelling (%)	Swelling (%)
$K^+ ==> H^+$	16	9
$K^+ ==> Na^+$	6	3.5

SBA resins	
Ion	Swelling (%)
$Cl^- ==> OH^-$	25-30
$Cl^- ==> SO_4^{2-}$	5
$Cl^- ==> NO_3^-$	-5

The weak acid and the weak base resins show a particular behavior due to their weak electrolyte nature. In fact, WAC in the H form or the WBA in the free base form are to a large extent non-ionized and consequently they are hydrophobic, the polymer chains collapse and the MHC is low. When they are converted to some other ionic form, then ionization becomes 100% and as the functional groups become hydrated, the resin becomes hydrophilic and swells considerably. For example, a WAC resin swells in the range 60-100% going from the H to the Na^+ form.

However, as the bigger counter-ion takes the place of a smaller one, the polymer matrix will apply a higher swelling pressure and even though the resin will expand, it will expel some of the free water. The higher the crosslinking density, the higher will be the internal swelling pressure and more "free" water will be expelled. The swelling therefore from one ionic form to another will depend on the crosslinking density. For example, a 8% DVB styrenic sulfonic resin swells by 16% going from the K^+ to the H^+ form while a 10% DVB resin swells by 9%. With resins with very high degree of crosslinking, the hydration of the ions may be incomplete with the result that swelling is very much suppressed and even it can be reversed since the size sequence of the naked ions is in the reverse order. Similar phenomena may be observed when the resin is found in very concentrated solutions where due to osmotic pressure difference the resin becomes dehydrated.

For resins with low crosslinking density where the free water represents a big part of the moisture content of the resin, the effect of the counter-ion valence on swelling becomes significant. In fact, it is the number of hydrated counter-ions that contributes to the moisture content. Since on the same equivalent basis, the number of divalent for example ions is half of those of monovalent, by converting a low crosslinked resin from the monovalent to the divalent form, the resin shrinks. In addition, since two functional groups are occupied by one divalent ion, the polymer chains may be less expanded due to a kind of a physical crosslinking effect. For example, a 8% DVB SAC resin shrinks by about 3% going from the Na^+ to the Ca^{2+} form. With a 6% DVB SAC the shrinking is about 6%.

With highly crosslinked resins, this may be reversed because there is not much free water and the divalent ions may be highly hydrated so that even with half the number, the resin may swell going from mono- to divalent form (Helfferich, 1962). For example, a 6% DVB SAC going from the Na^+ to the Mg^{2+} form shrinks by about 2% while a 16% DVB swells by about 3%.

When an IER comes in contact with an aqueous solution rather than with pure water, the resin swells less than in pure water because the osmotic pressure difference between the resin and the external solution is smaller. If the external solution is very concentrated, it can be that the resin shrinks as water is displaced from the resin to the external solution. This is seen for example when the resins are regenerated with concentrated acids or caustic solutions. It is also seen when resins treat concentrated sugar syrups. In these cases the "free" water is considerably reduced and the kinetics of ion exchange is reduced as well. Also, the properties that characterize ion exchange resins, such as moisture content, true wet density, expansion and pressure drop curves or swelling from one ionic form to another, as usually provided by the resin manufacturers in product data sheets, refer to resins in equilibrium with pure water. When the resin is in equilibrium with a solution, then these properties may have different values.

Total exchange capacity

Total exchange capacity of a resin is the quantity of ionic functional groups contained a certain quantity of resin. It is expressed in two ways, the number of equivalents of functional groups per unit of dry weight of resin, called the weight capacity (Wt Cap), or the number of equivalents of functional groups in a given volume of resin, called the volume capacity (Vol Cap). As the "volume" in the definition of the Vol Cap, is meant the apparent wet settled volume that the resin occupies in a vertical cylindrical vessel (figure 3.1). The apparent volume includes the real, or true, resin beads volume plus the interstitial volume between beads. The interstitial volume divided by the apparent volume is called void fraction. In common practice, the apparent wet settled volume is measured by placing a given quantity of resin in a graduated cylinder filled with water and topped up with water a few centimeters above the resin. The cylinder with the resin is then tapped until the resin height in the cylinder remains constant. If working with a column, the apparent volume is measured by fluidizing the resin bed for about 15-20 minutes to an expansion of about 100% (see particle size, below), stopping the flow and allowing the resin to settle and then passing water downflow through the resin bed at a velocity of 20 m/h for 15 minutes and then measuring the resin bed height. The volume is then calculated from this height and the column cross sectional area.

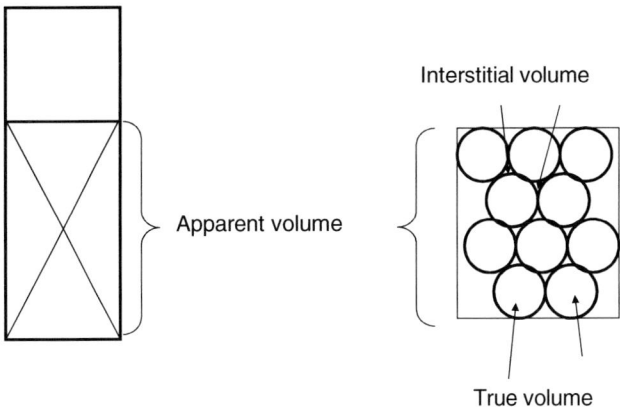

Figure 3.1   Resin volume

Although the volume capacity is a less precise measurement than the weight capacity, it is more practical to use the volume capacity because most use a certain volume rather than a certain weight of resin in practice. However, either measurements can be used as for example in the former Soviet Union countries end users continue to use the unit of weight for all IER handling.

Density

The different use of the "volume" and the "weight" of ion exchange resins necessitates different definitions of the density of a resin.
To begin with, we define the wet weight of a resin as the weight of a drained resin, that is, without the interstitial water, in equilibrium with air of 100% relative humidity. This is achieved by placing the resin covered with water in a Buchner and removing the water from the bottom of the resin by means of a vacuum pump. The air displacing the interstitial water is saturated with water vapor and the vacuum/air suction is maintained for 5 minutes. The crucible is then immediately wiped of all free water and weighed.

The following definitions are used:
- Apparent density: is the drained wet weight of a resin that gives a certain apparent volume in a vertical cylinder under water.
- Apparent volume: is the true resin volume divided by $(1-\varepsilon)$ where ε is the void fraction of the resin bed. It is measured as described above.

- Dry weight : is the drained, wet weight multiplied by the solids fraction (eq. 3.1)
- True wet density (TWD): is the drained wet weight of the resin that gives a certain true volume.
- Skeletal density: is the density of the dry polymer matrix of a resin. The skeletal density of a styrene-DVB copolymer for example is 1.066 g/ml.

The TWD is determined using a pycnometer or else, by measuring the terminal velocity of beads of a given size in a graduated cylinder filled with water and applying Stoke's law. As discussed above, the ionic form of the resin affects the true volume (the resin swells or shrinks), the moisture content of the resin, as well as its wet weight, thus affecting the true density. As an example, Amberlite™ IRA400UC, a SBA resin by Rohm and Haas Company, has the following TWD in various ionic forms (Rohm and Haas Co., 1980):

Form	*TWD (g/ml)*
Cl	1.09
HCO_3	1.11
SO_4	1.13
$UO_2(SO_4)_3$	1.32
$Fe(SO_4)_2$[1]	1.20

If we use the above definitions, we obtain the following relationship between volume and weight capacity:

Vol Cap = (Wt Cap)*(solids fraction)*(apparent density)   (3.2)

---

[1] Resin in equilibrium with a solution of 2.5 g Fe^{3+}/L and 22 g SO_4/L at pH=1.7

Volume capacity is usually expressed as equivalents per liter resin (eq/L_R). Weight capacity is usually expressed as equivalents per kilogram of dry resin (eq/kg). In both cases, the ionic form of the resin should be mentioned since the volume as well as the weight of the resin depends upon the ionic form.

Particle size

The particle size (the diameter) of the IER affects their performance in the various applications in a number of ways. In particular, the kinetics of ion exchange depend upon the bead size and this will be discussed in the next chapter. Particle size also affects the pressure drop across the resin bed and the degree of fluidization in a fluidized bed. In fact, the surface area of the resin beads depends directly upon the particle size of the resin and it is this surface area that is related to the kinetics, pressure drop or fluidization of a resin bed. For a given resin quantity, the smaller the particle size is, the bigger is the surface area of a given quantity of resin.

IER made by conventional suspension polymerization under agitation in a reactor have beads with wide particle size distribution, close to a Gaussian distribution. Often these resins are referred to as "Gaussian resins". IER made with some jetting technique for example, have a narrow particle size distribution and these resins are often referred to as "narrow particle size resins".

The particle size analysis of IER is traditionally given as the volume fraction that has a given size (diameter). Two parameters, the Effective Size (ES) and the Uniformity Coefficient (UC), are used to define the particle size distribution of a resin.

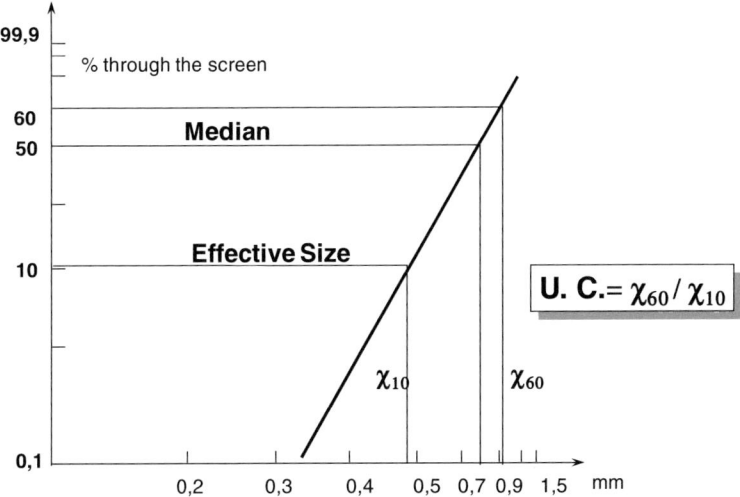

Figure 3.2 log-log plot of % of resin passing through a screen as a function of the aperture of the screens

After measuring the volume of each particle size across the full particle size of all the beads, a log-log plot is made of the cumulative volume percent of beads that are smaller than a certain diameter as a function of the particle diameter, in mm (figure 3.2).

The ES is defined as the sieve size, χ_{10}, corresponding to 10% of the resin volume passing through this sieve size. The UC is defined as the ratio of the size that corresponds to the 60% of the resin volume to the size at 10% of the resin volume, as seen in figure 3.2.

The average size is expressed as the median diameter, which is the sieve size at which 50% of the resin volume is smaller than that sieve (fig. 3.2).

There are however various other ways to give an average particle size of a resin, such as the arithmetic mean (AM), the geometric mean (GM) or the Harmonic Mean (HM) diameter. The relationship among these averages is:

$$AM \geq GM \geq HM$$

The equal sign applies to monodisperse distributions. Even with these averages, there are other definitions such as number mean diameter, based on the number of resin beads of a given diameter, volume mean diameter, based on the volume of beads that have a given diameter, etc. In ion exchange resins the volume average is traditionally used. In particular, the Harmonic Mean Size (HMS) is frequently used because it gives an average value that is related to properties such as pressure drop. HMS is defined as:

$$HMS = \frac{1}{\Sigma (x_i/D_i)} \qquad (3.4)$$

where x_i is the volume fraction that has diameter D_i.

The HMS of a resin is more related to the properties that depend on the surface of the beads, than the other means, such as the arithmetic mean. For example, the pressure drop ΔP across a resin bed, well classified, when calculated using the Ergun equation (see next topic), from each volume fraction that corresponds to a size of the beads and add up to get the total pressure drop, is close to the ΔP calculated directly from the HMS.

This is illustrated in Table 3.1

Table 3.1

Scree size (μ)	Volume fraction	Pressure drop (kg/cm2/m)
250-300	0	0,000
300-355	0,006	0,004
355-400	0,007	0,004
400-500	0,049	0,018
500-600	0,132	0,032
600-710	0,188	0,032
710-800	0,157	0,020
800-850	0,085	0,009
850-1000	0,3	0,027
1000-1250	0,076	0,005
Sum		0.151

HMS = 0.724 mm
AM= 0.771 mm
ΔP from HMS   0.141 kg/cm2/m
ΔP from AM   0.125 kg/cm2/m

Pressure drop

The pressure drop (ΔP) through a bed of resin, depends on one hand upon the operating conditions, mainly the linear velocity and the viscosity of the liquid, and on the other hand the resin, mainly the particle size and the void fraction (the interstitial volume).

Linear velocity, or superficial velocity, is the volumetric flow rate divided by the empty column cross sectional area. Pressure drop is expressed as pressure per meter of resin bed in units of kg/cm^2, bars or kilopascals, kPa (100 kPa equals 1 bar, 1 kg/cm^2 equals 0.98 bar).

When the liquid flows through a resin bed, it passes through the interstitial space between the resin beads. Most pressure drop problems encountered (pressure drop becomes too high) result from a decrease of the interstitial space between the resin beads. This is typically due to either suspended matter or resin fragments which plug the openings between resin beads. If pressure drop is excessive, resin beads can deform and this too will partially plug or decrease the interstitial space.
In many instances resin beads dot not deform unless their moisture content is high (in the range 60% or higher for styrenic resins, less for acrylic resins), flow rates are fast and deep resin beds. Pressure drop can be calculated using the Ergun equation:

$$\Delta P/L = 4.96*10^{-4} * \frac{(1-\varepsilon)^2 * \eta * v}{\varepsilon^3 \; D^2} + 1.38*10^{-6} * \frac{(1-\varepsilon)}{\varepsilon^3} * \frac{\rho * v^2}{D}$$

where:
$\Delta P/L$ = pressure drop in kg/cm^2/m resin
ε = void fraction
η = fluid viscosity, cp
ρ = fluid density, g/cm^3
v = linear velocity, m/h
D = particle diameter, mm

The Ergun equation applies to both, laminar and turbulent flow through the resin packed bed. In fact, the first term prevails in laminar flow while the second terms prevails in turbulent flow. The Reynolds number when water flows at a superficial velocity up to 40 m/h at ambient temperature through a packed bed of a resin with particle size 0.6 mm and a void fraction of 0.38 is less than 10 and hence, the flow is laminar. In this range, the pressure drop is proportional to the superficial velocity and independent of the fluid density.

In order to compare the calculated values using the Ergun equation with real values, or to predict the pressure drop of a resin in a given application, one has to define the particle diameter and to determine the void fraction of the resin because this factor affects greatly the pressure drop and because it is not, strictly speaking, a property of the resin, as it is the particle size for example.

As particle size the HMS can be used, as discussed in the previous section. The void fraction can be determined experimentally by injecting a substance that is not absorbed by the resin and measuring the volume of the effluent when this substance exits the column, or else, by draining the column to the top of the resin and then push slightly with air the interstitial water out of the column and measure the volume of the exiting water.

The interstitial volume depends upon the linear velocity of the fluid, the resin moisture (if the resin beads are deformable or not), and the packing of the resin achieved after a backwash. The closest packing of equal beads possible has a void fraction of ≈ 0.26. Usually, the void fraction of IER is found to be between 0.36 and 0.40. A standard styrenic resin with normal particle size distribution, after a classification with a backwash, would give a void fraction of 0.37-0.38. In mixed condition, the same resin would give a void fraction of about 0.36.

Figure 3.3 illustrates calculated pressure drop using the Ergun equation compared to real values with a small particle size resin.

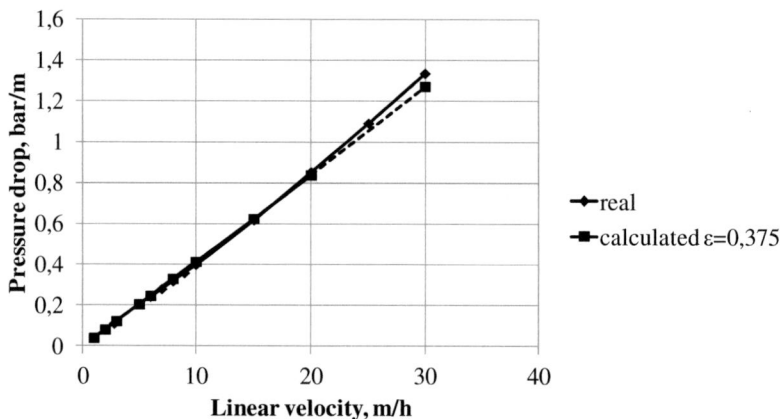

Figure 3.3 Real vs calculated ΔP using the Ergun equation

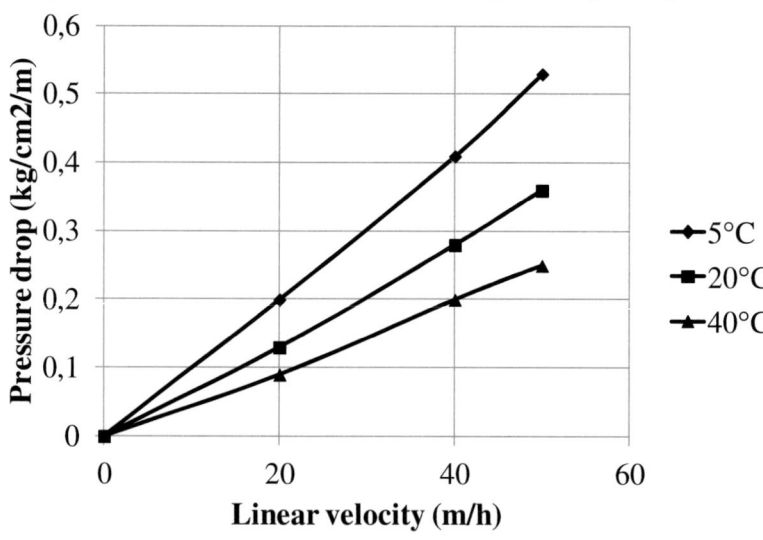

Figure 3.4 Effect of temperature on pressure drop for a styrenic resin in water having an HMS of 750 μ

Figures 3.4 and 3.5 give the pressure drop of a styrenic resin in water as a function of the linear velocity of the liquid and for various temperatures and resin particle size.

The dependence of pressure drop on temperature is due to the change in the fluid characteristics (mainly the viscosity).

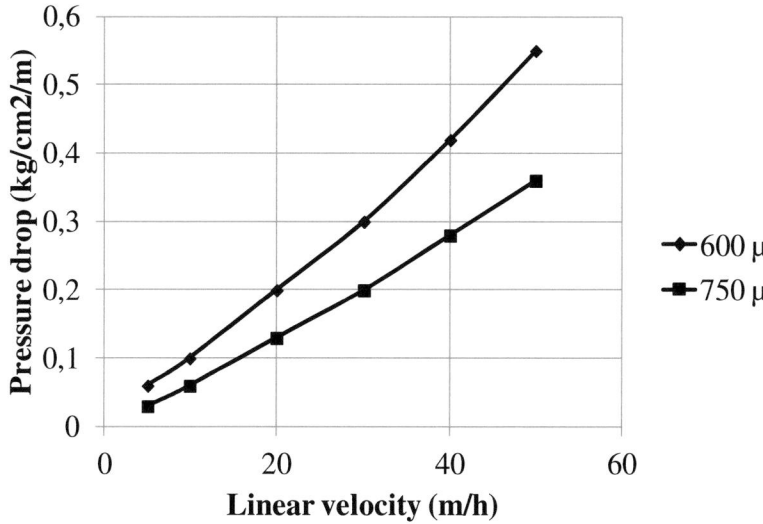

Figure 3.5 Effect of particle size on pressure drop for a styrenic resin in water at 20°C

Resin fluidization

When a liquid is introduced into a column containing an ion exchange resin in upflow direction, at the beginning, as long as the velocity of the liquid is slow, the resin bed does not move. Progressively, as the velocity of the liquid increases, a certain velocity is reached, called minimum fluidization velocity V_{mfv}, where the upper part of the resin bed will fluidize while the lower part will remain still. As the liquid velocity increases further the resin bed will expand until it reaches a point where the whole resin bed will fluidize. The lighter (that is, the smaller) beads will tend to go to the upper part while the heavier (the bigger) beads will remain at the lower part. By increasing further the liquid velocity, the fluidized bed will continue to expand. When the liquid velocity attains the final terminal velocity of the beads found at the top of the expanded resin bed (Stoke's law), these beads will be carried out of the column by the liquid.

The degree of resin fluidization depends on the linear velocity of the upflow flowing fluid, the fluid viscosity, the particle size of the resin and the density difference between resin and fluid. Since temperature affects both viscosity and density, it will also affect the degree of fluidization of the resin. The unwanted particles of smaller size than the size of the beads at the top of the bed are carried away by the out flowing liquid. In this way, any fine particles or resin fragments can be removed from the resin bed in order to ensure a low pressure drop (fig. 3.6).
The height of the fluidized bed in a cylindrical column minus the initial bed height divided by the initial bed height of the resin gives the bed expansion, usually expressed as percent:

$$\% \text{ Bed expansion} = \frac{\text{(final bed height} - \text{initial bed height)}}{\text{(initial bed height)}} * 100$$

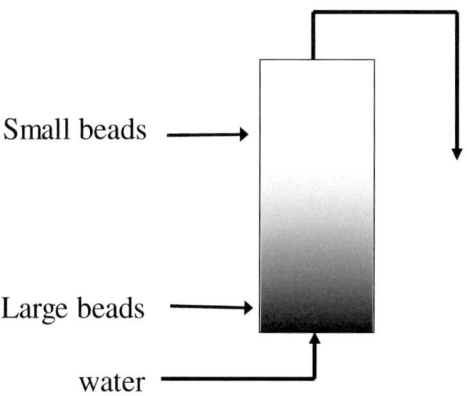

Figure 3.6   Resin fluidization and classification

After fluidization has reached a steady state and the fluidized resin beads remain at their respective height, we say that the resin has been "classified". The migration of the beads to their steady state position through the fluidized resin bed takes place easily provided that the resin bed expansion is sufficient. Otherwise, small beads, or fragments, remain trapped by larger beads and to reach eventually their steady state position they will require a very long backwash time. Usually, for an efficient backwash, the resin bed should be expanded by at least 60% and preferably more.

After decantation, a classified polydisperse resin occupies approximately 3-5% more volume than a mixed resin. Since the true resin volume does not change, this means that the void fraction of a mixed resin is smaller than for a classified resin.

It is important to have expansion curves as a function of linear velocity at different temperatures for a given liquid depending on the application of the resin. These are used in backwash operations where the removal of suspended matter from the resin bed is required. Failure to observe these parameters could result in inadvertent loss of resin as it is carried away from the column by the liquid.

Figures 3.7 and 3.8 show expansion curves of a SBA resin in the Cl form and in water as a function of the liquid linear velocity at various temperatures and resin particle size.

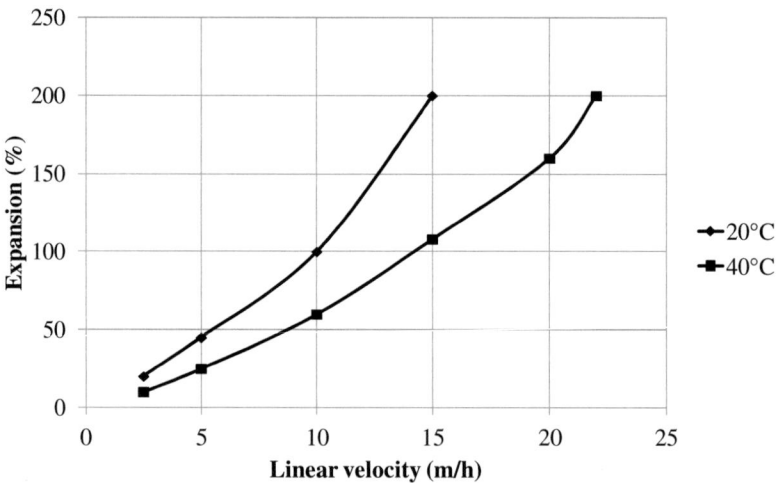

Figure 3.7 Effect of temperature on resin fluidization for a SBA resin in Cl form in water

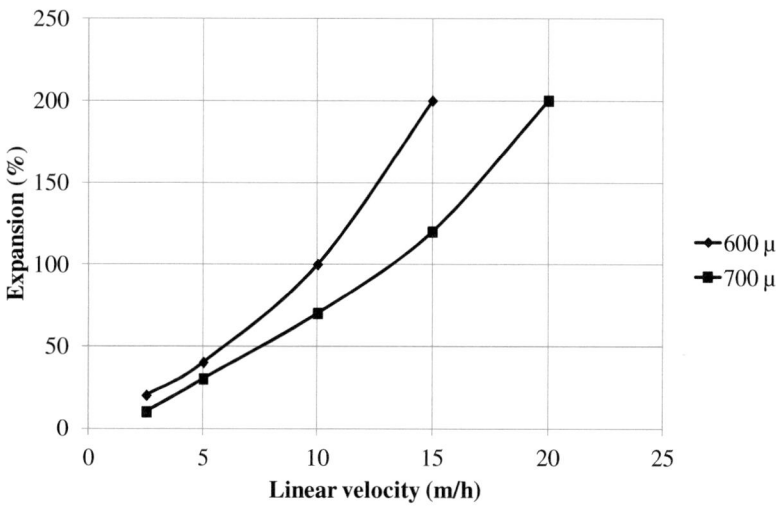

Figure 3.8 Effect of particle size on resin fluidization for a SBA resin in Cl form at 20°C in water

Porosity, pore size, pore size distribution and internal surface area

This paragraph applies to MR resins and polymeric adsorbents. Gel type resins, by definition, do not have (macro)pores.
The internal surface area of MR resins is determined by an inert gas adsorption technique. A dry, clean quantity of resin is placed in a vacuum chamber where an inert gas, such as nitrogen, is introduced at very low temperatures (liquid nitrogen) and at various pressures. The amount of gas adsorbed by the resin is determined by the variation in the pressure. In this way, adsorption

and desorption isotherms are constructed. By knowing the surface area of a molecule of the gas, N_2 in this case has 16.2 $Å^2$, and using an appropriate model, the total surface area per dry weight of adsorbent (specific surface area) of the resin can be determined. The most widely used model is the BET model (Brunauer et al, 1938). The pore size and pore size distribution can be determined from adsorption and desorption isotherms and using an appropriate model (the Kelvin model) which assumes certain shape and structure of the pores. The porosity of the adsorbent, P, which is the volume of the pores per unit weight or volume of the adsorbent, can be determined from the true wet density and the polymer density of the adsorbent. It can also be determined from the nitrogen adsorption during surface area determination and some BET analyzers give porosity along with specific surface areas.

As mentioned in Chapter 2, the internal surface area of MR resins is the surface area of the microbeads that are formed during polymerization. The size of these microbeads is a function of the polymerization conditions: initiator concentration, monomer composition (% DVB), temperature, porogen type and quantity. The smaller the microbeads, the bigger the specific surface area. The porosity P of the resin depends on the quantity of the porogen in the mixture of the reagents.

It is possible to calculate an average pore size (average pore diameter) from the specific surface area, S, and the porosity per dry weight, P, by assuming a cylindrical shape of the pores and using the expression:

$$\text{Average pore diameter} = 40000 * P / S \quad (3.5)$$

where the average pore diameter is obtained in Angstroms (Å), and where P is the porosity in ml/g and S the specific surface area in m^2/g.

Understandably, the use of this average pore diameter is of limited use since it does not give the distribution of sizes.

In general, the synthetic adsorbents are made under conditiond such that a high surface area is obtained at a given pore size and pore size distribution. Conventional ion exchange resins have much lower surface area than the adsorbents. As an indication, conventional IER have a specific surface are of about 30-50 m2/g and an average pore diameter of about 300 Å, while synthetic adsorbents have 500-1000 m2/g and 50-150 Å respectively.

The moisture holding capacity of a MR resin is the total quantity of water, in grams, included in the macropores and in the polymer phase per hundred grams of drained, wet resin. An ion or a molecule fixed or adsorbed by an MR resin penetrates into the resin beads mainly through the macropores from which they diffuse into the polymer phase of the microbeads. In cases where these ions or molecules are voluminous and diffusion through the gel resin is slow, MR resins offer better kinetics compared to gel type resins, as discussed in the next chapter.

In order that a molecule is adsorbed on an adsorbent, it should be smaller than the size of the pores through which these molecules diffuse into the adsorbent. Adsorbents that have rather large pores near the external bead surface while having smaller pores in the interior, show good kinetics of adsorption and high adsorbing capacity.

Physical and chemical stability of IER

Physical degradation and breakdown of IER can occur for the following reasons:
- volume changes due to swelling and shrinking
- mechanical impact on external surfaces such as pump impellers etc.
- existing microcracks or weak points on the resin beads
- excessive and rapid thermal shocks (during heating or cooling)
- precipitation inside resin beads
- rapid freezing of internal water
- attrition due to abrasion by suspended matter in the solution, for example in Resin-in-pulp operations

Although there exist general rules for expecting resin attrition, for example high swelling and shrinking can cause resin breakage, a lot depends on the manufacturing formulations of the ion exchangers. Each ion exchange product has its own resistance to breakdown, for example, two different products, even of the same type, that swell/shrink to the same extent, do not necessarily break to the same extent. In some applications it is difficult to change the operating conditions. In these cases, resins should be evaluated beforehand in order to determine the most suitable resin. In other cases wherever possible, one should choose the operating conditions to ensure long resin life.

A frequent cause of resin breakdown is the rapid and high volume changes that the resin undergoes during service or regeneration. When the resin undergoes a change in ionic form, it can swell or shrink, as previously described. If this volume change is significant, greater than 20-25 % for example, and if in addition to this, this volume change takes place rapidly, for example when the solution contains high concentration of the ion to be exchanged, then the stresses developed in the resin under these conditions are very severe and can result in breakage of the resin bead. This volume change is called sometimes " osmotic shock". The resistance to breakdown due to this volume changes, for a given product, depends on the degree of swelling and shrinking of the resin. The higher the swelling and concentration variations, the more the resin beads tend to break. Also, beads that have already weak points (microcracks etc) from the manufacturing stage show low resistance to breakdown.

As discussed previously, the higher the crosslinking density of gel-type resins, the less the resin swells and shrinks upon changes from one form to another or upon changes in ionic environment. One way therefore to improve the physical stability of a resin is to use resins with higher crosslinking density, up to a certain extent. For example, it is a general experience that a gel type SAC resin having 10% DVB level resists better than a 8% DVB resin. It should be noted that for macroporous (macroreticular) resins, there is no simple correlation between crosslinking density and swelling. There exist macroporous resins of the same chemical structure and the same total volume exchange capacity that swell to a different degree going from one form to another. Macroporous resins have, due to their physical structure, improved resistance to breakdown caused by volume changes compared to gel type resins.

In the case of anion exchangers, especially in the case of WBA resins that swell considerably, the physical stability may be improved by increasing the secondary crosslinking through methylene bridging (Girsh et al 2001). A different way to improve the stability of WBA resins is by varying the content of strong base groups. Since the volume changes of the weak and strong base groups from the exhausted (Cl) form to the regenerated (OH) form are in opposite directions, with the result that the strong base groups swell while the weak base groups shrink, by varying the content of strong base groups, the degree of swelling changes and so does the resistance to breakdown.

In practice, whenever feasible, use low solution concentrations so that the conversion of the resin from one form to the other takes place slowly and in this way breakdown is significantly reduced.

In general, large size beads of a given type of product, tend to break easier than small size beads. The comparison should be made for the same type of product (similar manufacturing process).

Another cause of breakdown is when the resin is transferred either from one column to another or when it is charged into a column, by means of a centrifugal pump. It is strongly recommended that transfer of resin be made either hydraulically, by fluidizing and pushing the resin with water pressure or by use of a hydro ejector.

A rapid thermal shock can also damage the resins. A rule of thumb is a rate of temperature change across a resin bed should not exceed 3°C per minute. This rule holds for new resins in good condition.

Sometimes when the external temperatures drop and the resins are exposed to this temperature, for example during resin storage in countries where the winter temperatures drop down to many degrees below zero, the interstitial water and even the internal (free) water in the resin can freeze. Under normal conditions, the interstitial water of a resin in its original packaging has been drained, so only a small part remains. The internal water contains a certain concentration of ions which decrease the freezing point of this water to at least several degrees below zero. Nevertheless, if the free water in the resin freezes, care should be taken to progressively thaw the resin in order to prevent resin breakage and damage.

Abrasion of resin by suspended matter in the solution is an important parameter in applications where the resin is mixed with high suspended solids solutions (pulps) in a batch contactor under agitation. There are tests developed to test the resistance of the resins in these kind of applications. One of these, called ball mill test, consists in allowing a certain quantity of resin to roll together with a certain quantity of spheres over a period of many hours or days, after which the particle size of the resin is measured and compared to the initial particle size. A similar test, which simulates real application conditions, allows the resin to be agitated with a certain quantity of the high suspended solids liquid for a given period of time, and again compares the particle sizes before and after the test.

Chemical degradation of resins can be essentially due to two causes:
- thermal degradation
- oxidation

Sulfonic resins in the H^+ form can lose capacity by the reaction of figure 3.9 where the sulfonic groups are split and form H_2SO_4 leaving the resin with only the styrene unit. However, for most of the common SAC resins, for temperatures up to 120°C, the rate of this reaction is so slow that we can say that sulfonic resins are stable up to these temperatures. Above 140°C, these resins loose capacity rapidly.

This reaction is catalyzed by H^+ and therefore the rate of thermal degradation also depends on the pH. Consequently, sulfonic resins in other ionic forms such as Na^+, K^+ etc, show increased thermal stability.

When the benzene ring contains electron-repelling groups, like alkyl groups, the sulfonic group splits off more easily. On the contrary, when the benzene ring contains electron-attracting groups such as -Cl or -NO_2, the thermal stability of the resin increases. In fact, sulfonic resins containing -Cl groups in order to improve their thermal stability have been developed and these are used in heterogeneous catalysis as catalysts and are able to operate at temperatures exceeding 150°C.

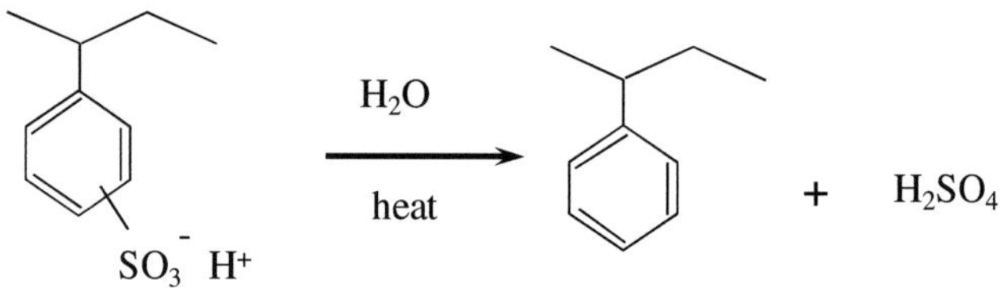

Figure 3.9 Thermal degradation of sulfonic resins

SBA resins are the least thermally stable and the thermal stability depends on the ionic form. SBA resins in OH⁻ form are much less stable. They convert to weak base resins by the Hoffman degradation as shown in figure 3.10 for type 1 SBA resins.

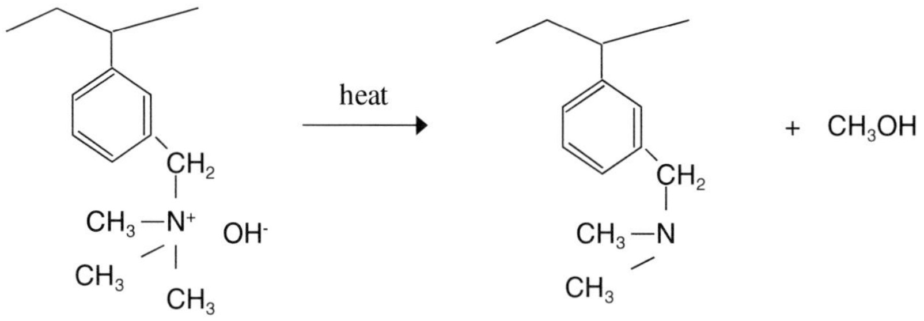

Figure 3.10 Hoffman degradation

Another mechanism is the degradation where the whole functional group is lost:

$$P\text{-}CH_2\text{-}N^+(CH_3)_3 OH^- \longrightarrow P\text{-}CH_2OH + N(CH_3)_3$$

where P- denotes the polymer backbone.
Type 1 SBA resins in OH⁻ form can be used up to 50°C before they start loosing strong base capacity. Type 2 resins in OH⁻ form and acrylic SBA resins in OH⁻ form are stable only up to 35°C. Acrylic SBA resins in Cl⁻ form are stable only to about 50°C. In other ionic forms, such as Cl⁻ form, the Type 1 and Type 2 SBA resins can be used up to 80°C and 60°C respectively.
Under elevated temperature conditions, acrylic resins undergo degradation through the hydrolysis of the amide groups (fig. 3.11).

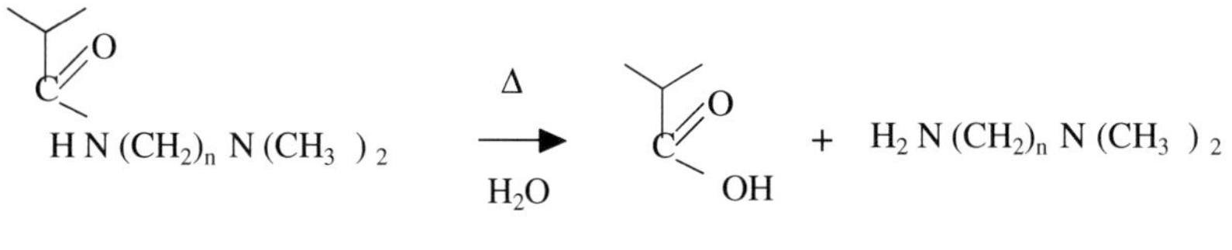

Figure 3.11 Thermal degradation of acrylic resins

A new type of styrenic SBA exchange resins have been synthesized having long aliphatic chains between the benzene ring and the quaternary amine group (fig. 3.12). These resins are the Diaion® XSA series by Mitsubishi Chemical Corporation (Masuda *et al*, 2000). These resins can stand up to 140°C in the Cl⁻ form and up to 80°C in the OH⁻ form.

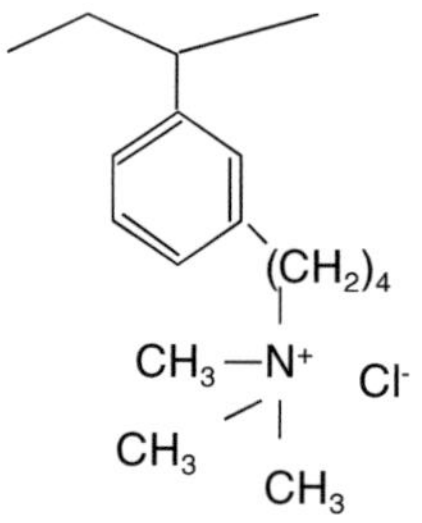

Figure 3.12 Diaion® XSA series of SBA resins

Exposing the IER to oxidants such as Cl_2 or H_2O_2 or even O_2 (in the presence of a catalyst such as Fe etc) over long periods of time and at oxidant concentrations above 0.1 to 1.0 ppm depending on the resin type, will degrade the resins by cutting the polymer chains (fig. 3.13a) and/or splitting off the anion exchange groups (fig. 3.13b).

These reactions are exothermic. In the presence of strong oxidants at high concentration, the oxidation reactions can become uncontrollable with extremely dangerous consequences. For example, concentrated HNO_3 has caused explosions of columns containing IER when this was accidentally introduced to the resin bed. Similar dangers are faced with chromic acid, concentrated H_2O_2 and other strong oxidants.

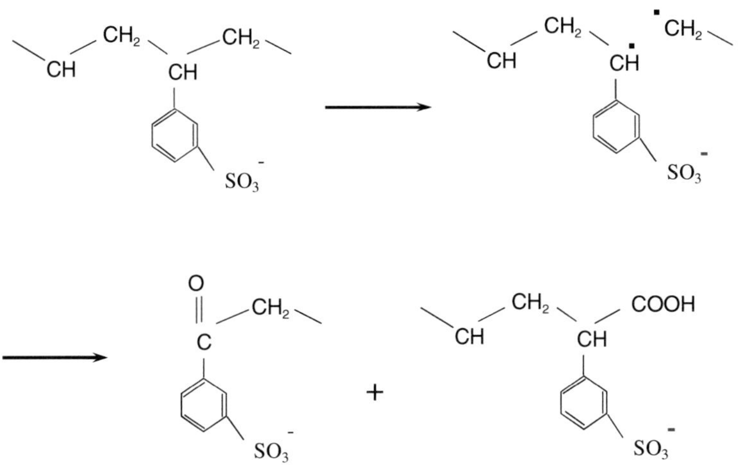

Figure 3.13 Oxidation of IER: (a) chain cut (de-crosslinking); (b) loss of capacity

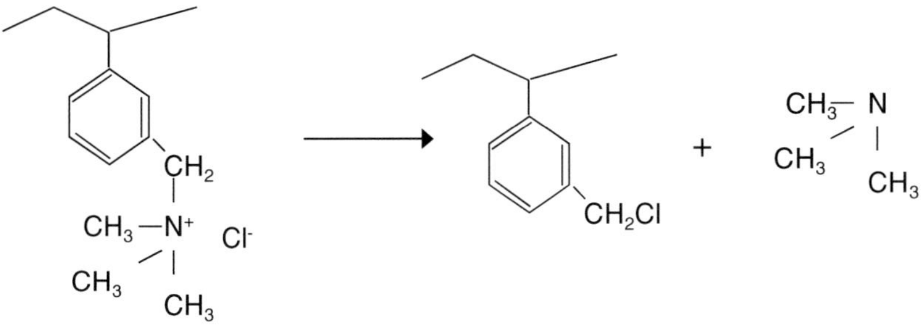

Dissociation of weak electrolyte resins

As discussed earlier, the functional groups of the resins can be strong or weak electrolytes. This results in different ion exchange properties between strong and weak electrolyte resins.
Strong electrolyte resins have functional groups 100% ionized or dissociated in all circumstances. Weak electrolyte resins on the other hand, in the acid form for the WAC resins or the free base form for the WBA resins, have functional groups which are only partially dissociated.
Consider WAC exchange resins in the H form. The carboxylic groups will be dissociated according to the reaction:

$$\text{R-COOH} \leftrightarrows \text{R-COO}^- + \text{H}^+ \qquad (3.6)$$

Where R indicates the resin matrix, with a dissociation constant given by:

$$K_a = \frac{[\text{R-COO}^-][\text{H}^+]}{[\text{R-COOH}]} \qquad (3.7)$$

where [....] indicate concentrations inside the resin beads. If the $[\text{H}^+]$ concentration is very high, reaction (3.6) will go to the left, thus at low pH, WAC resins do not ionize. Equation (3.7) can be written as:

$$\text{pH} = \text{pKa} + \log \frac{[\text{R-COO-}]}{[\text{RCOOH}]}$$

Thus at a pH = pKa the carboxylic resin is 50% dissociated. At a pH one unit above the pKa, over 90% of the carboxylic groups will be ionized while at a pH one unit lower than the pKa, over 90% of the carboxylic groups will be non-ionized.

Now consider the case where a carboxylic resin is placed in contact with a solution containing a salt M^+X^-. The exchange reaction is:

$$RCOOH \leftrightarrows R\text{-}COO^- + H^+ \qquad (3.8)$$

$$R\text{-}COO^-H^+ + M^+X^- \leftrightarrows R\text{-}COO^-M^+ + HX \qquad (3.9)$$

$$HX \leftrightarrows H^+ + X^- \qquad (3.10)$$

If acid HX is stronger than the carboxylic group of the resin, in other words if H^+ prefers the $-COO^-$ group from the X^- ion, then reaction (3.9) will go to the left and the resin will not significantly fix M^+ ions. In this case we say that the carboxylic resin does not salt split. In order that reaction (3.9) goes to the right, the acid HX should be weaker than the $-COOH$, or that the dissociation constant Ka (resin) is bigger than the dissociation constant Ka (acid HX), or

$$pKa \text{ (resin)} < pKa \text{ (acid HX)} \qquad (3.11)$$

Polyacrylic acid has a pKa of about 5 while polymethacrylic acid has a pKa of about 6. Carbonic acid has a pKa_1 of 6.34. This means that a polyacrylic acid resin in the H form can fix Na^+ from a $NaHCO_3$ solution and replace it with H^+. Since however the two pKa values, that of the resin and that of the carbon-

71

ic acid, are not very different, the resin will fix only partially Na^+ in exchange for H^+ and certain part will remain in the H form.

WBA exchangers react in the same way:

$$R\text{-}CH_2N(CH_3)_2 + H_2O \leftrightarrows R\text{-}CH_2NH^+(CH_3)_2 + OH^-$$
(3.12)

with
$$K_b = \frac{[R\text{-}CH_2NH^+(CH_3)_2]\,[OH^-]}{[R\text{-}CH_2N(CH_3)_2]}$$

Here we also have $K_a = K_w / K_b$ where K_w is the dissociation constant of water which is equal to 10^{-14}. Therefore, $pK_a = 14 - pK_b$. If the [OH$^-$] concentration is very high, the reaction (3.12) goes to the left : WBA resins at high pH do not ionize. The pKa of a styrenic WBA resin is about 8 to 8.5, the pKa of an acrylic WBA resin is about 9. In other words, the acrylic WBA resins are a little stronger bases than the styrenic WBA resins. A WBA exchanger in water will therefore be mainly in the non-dissociated form. We say that the resin is found in the free base form.

Assume the case of a WBA resin treating a solution containing an acid. (This is the case in water treatment where a SAC resin in the H^+ form precedes the WBA resin. The SAC resin exchanges cations for H^+ ions turning the water acidic). For example:

$$R\text{-}CH_2N(CH_3)_2 + H^+X^- \leftrightarrows R\text{-}CH_2NH^+(CH_3)_2 + X^-$$
(3.13)

If the proton H$^+$ prefers the X$^-$ over the functional group of the resin, in other words if the acid HX is weaker acid than the resin, or

$$pKa\ (HX) > pKa\ (resin) \qquad (3.14)$$

then the reaction (3.13) will go to the left, hence, according to (3.14), WBA resins do not fix very weak acids such as silicic acid, H$_2$SiO$_3$, which has has a pKa$_1$ of 10.

In a similar way as with carboxylic resins, WBA resins do not salt split:

$$R\text{-}CH_2N(CH_3)_2 + Na^+X^- \underset{}{\overset{H_2O}{\rightleftarrows}} R\text{-}CH_2NH^+(CH_3)_2 + Na^+OH^- \qquad (3.15)$$

The NaOH formed in this reaction raises the pH too high for the weak base group to ionize. The reaction (3.15) will go to the left.

4. Ion exchange equilibrium and kinetics

Ion exchange equilibrium

IER contain, as we saw, ionic functional groups of a given charge, balanced by an equivalent amount of counter ions of opposite charge. In contact with a dilute solution of an electrolyte, the ions of the electrolyte with the same charge as the functional groups of the resin are called co-ions while the ions with opposite charge are called counter-ions.

When the resin comes in contact with a dilute solution of an electrolyte with the same counter ions as the resin but different co-ions, the concentration of the counter ions in solution is lower than that in the resin while the concentration of the co-ions in solution is higher than that in the resin.

Take a SBA in the Cl^- form as an example (figure 4.1) and in contact with a dilute solution of NaCl. Because of the concentration difference between resin and solution, counter ions (here : Cl^-) will tend to migrate from the resin into the solution phase while co-ions (here : Na⁺) will tend to migrate from the solution to the resin phase. However, as the counter ions diffuse out of the resin into the solution and co-ions from the solution into the resin, they create a positive charge in the resin and a negative charge in the solution. The established electric potential between the two phases, called Donnan potential (Helfferich 1962, p. 134), will pull back Cl⁻ from the solution into the positively charged resin and the Na⁺ from the resin into the negatively charged solution.

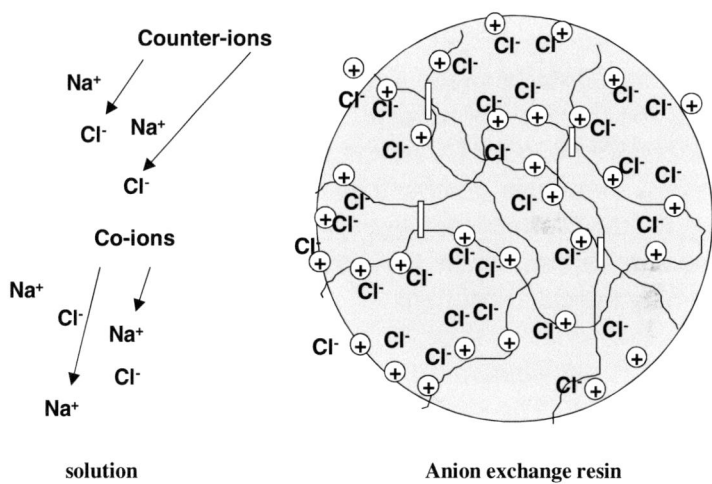

Figure 4.1 Anion exchange resin in contact with a solution of NaCl

There will then be an equilibrium, called Donnan equilibrium, where the concentration difference will be balanced by the electrical field. Thus, the Donnan potential will prevent the positively charged co-ions (the Na⁺) from entering the positively charged resin and because of electroneutrality, the electrolyte (here NaCl) will be prevented from entering into the resin. This phenomenon is called ion exclusion. The potential between resin and external solution at equilibrium is given by the Nernst equation:

$$E = (RT/zF) * \ln(c_{out}/c_{in})$$

where :
- R=gas const=2 cal mol^{-1}°K^{-1}
- T=temp=273+°C
- z=valence=+1 for Na$^+$ and -1 for Cl$^-$
- F=Faraday const=23000 cal V^{-1}mol^{-1}
- c= concentration in molarity, M

For each ion there is the corresponding potential E, for example one for Na$^+$ and one for Cl$^-$ in our case. When the electrical potential E_{Na+} for the Na$^+$ ions will be equal to the electrical potential E_{Cl-} for the Cl$^-$ ions, then there will be an equilibrium, called Donnan equilibrium, where the concentration difference will be balanced by the electrical field.

Analogous is the situation of a SAC resin in the Na$^+$ form in a solution of NaCl. The negatively charged resin will prevent the co-ions (Cl$^-$) from entering in the resin and due to the electroneutrality, NaCl will not migrate into the resin.

The Donnan potential required to balance the concentration difference when the counter ion has high valence is smaller because the force acting on the ion is proportional to its valence. For example, a divalent ion would need half the potential to balance the concentration difference than a monovalent ion.

Ion exclusion is less efficient with counter ions of high valency and it is more efficient with counter ions of low valency. Also, at a given Donnan potential, *co-ions with high valency are more strongly excluded than co-ions of low valency*. For example, a SAC resin will prevent Na_2SO_4 more strongly from entering the resin (here Na^+ is the counter ion and SO_4^{2-} the co-ion) than NaCl (Na^+ is the counter ion and Cl^- the co-ion). On the other hand, a SBA resin will exclude more strongly NaCl (here Na^+ is the co-ion and Cl^- the counter ion) than Na_2SO_4 (Na^+ is the co-ion and SO_4^{2-} the counter ion). Similarly, $CaCl_2$ will be more excluded by a SBA resin than NaCl (Ca^{2+} and Na^+ are co-ions).

The greater the concentration difference between the ion exchanger and the solution, the higher is the Donnan potential and the more efficient the ion exclusion. A high total capacity of the resin and a high crosslinking density means high concentration of the counter ion in the resin and for a given counter ion concentration in the solution will tend to increase the Donnan potential and therefore the ion exclusion is increased. Similarly, the Donnan potential and therefore ion exclusion increases when the solution concentration decreases.

Interactions between counter ions and co-ions can affect the Donnan potential and therefore the ion exclusion. A typical case is the complex formation whereby the Donnan potential is decreased. For example, in a solution of $ZnCl_2$ in contact with a SBA resin in the Cl form, the co-ion Zn^{2+} and therefore $ZnCl_2$, is excluded from the resin. However, the formed complex $ZnCl_4^{2-}$ can migrate into the resin.

If the counter-ions in the external solution are different from those in the resin, then the external counter-ions enter into the resin and exchange with those inside. The composition of the external solution will depend then on the selectivity coefficient of the resin for these ions, as it is discussed below.

If there is a neutral molecule in solution, then the concentration difference between resin and solution will be leveled off by diffusion of the neutral molecules into the resin. Therefore, neutral molecules are not excluded from the resin and it will distribute itself freely between resin and solution.
This is the basis of the ion exclusion chromatography where a neutral molecule, for example sugars, glycerin etc, in solution can be separated from salts present in the solution.

Let us take now an IER containing a given counter ion in contact with a solution containing an electrolyte with different counter ion. In this case, there will be an exchange between the counter ion initially in the resin and the counter ion initially in solution. In general, this exchange is reversible, it can take place when the counter ion in the resin is for example A and the counter ion in solution is B, or when the counter ion in the resin is B and that in solution A. The extent to which the exchange of the counter ion on the resin for the ions in solution will take place depends upon the preference of the resin for one or the other ion. This preference is expressed as a selectivity coefficient $K_{i/j}$ or a separation factor $a_{i/j}$ for a binary exchange. For a multicomponent ion exchange, it is assumed that there are no interferences of one ion on the exchange of another one, so the same selectivity coefficient or separation factor is used.
Consider the case of a SAC resin in the H^+ form (that is, the counter ion is a H^+) that comes in contact with an aqueous solution of NaCl. An exchange takes place of the H^+ for the Na^+. After equilibrium has been established, the IER, initially in the H^+ form, will be found in a form partially H^+ and partially Na^+. Similarly, the solution, initially NaCl, will contain both, NaCl and HCl. The ion exchange reaction is described by eq. (4.1):

$$R\text{-}H + Na^+Cl^- \leftrightarrows R\text{-}Na + H^+Cl^- \qquad (4.1)$$

where R represents the polymer matrix of the resin. Applying the law of mass action, this equilibrium condition is described with the equilibrium constant (4.2):

$$K_{Na/H} = \frac{\{R\text{-}Na\}\{H^+\}}{\{R\text{-}H\}\{Na^+\}} \qquad (4.2)$$

where $K_{Na/H}$ is the equilibrium constant and { } denote the activities of H^+ and Na^+ on the resin and in solution after equilibrium. For simplicity, if we use concentrations instead of activities, eq. (4.2) becomes:

$$K_{Na/H} = \frac{(R\text{-}Na)(H^+)}{(R\text{-}H)(Na^+)} = \frac{q_{Na}\, c_H}{q_H\, c_{Na}} \qquad (4.3)$$

where $K_{Na/H}$ is the selectivity coefficient and ( ) denote concentrations in equivalents per liter (eq/L) or equivalents per liter resin (eq/L_R).

q_{Na} = concentration of Na in resin in eq/L_R
c_H = concentration of H in solution in eq/L
q_H = concentration of H in resin in eq/L_R
c_{Na} = concentration of Na in solution in eq/L

Alternatively, the use of the separation factor, $\alpha_{i/j}$, defined as:

$$\alpha_{i/j} = \frac{\text{distribution of ion i between phases}}{\text{distribution of ion j between phases}} = \frac{y_i/x_i}{y_j/x_j} \qquad (4.4)$$

is more practical. Here :

x_i is the equivalent fraction of ion i in solution and
y_i is the equivalent fraction of ion i in the resin.

$$x_i = c_i / C \quad \text{and} \quad y_i = q_i / Q \qquad (4.5)$$

Where c_i is the concentration of ion i in solution in eq/L and C the total concentration of all ions in solution in eq/L. q_i is the concentration of ion i on the resin in eq/L$_R$ and Q is the total capacity of the resin in eq/L$_R$.

If the resin prefers ion i over ion j we call this a favorable equilibrium and $a_{i/j} > 1$. In the opposite case we call this an unfavorable equilibrium and $a_{i/j} < 1$.

Using eqs (4.4) and (4.5) we have for the above example of the H$^+$/Na$^+$ exchange :

$$\alpha_{Na/H} = \frac{q_{Na} \, c_H}{q_H \, c_{Na}} = K_{Na/H} \qquad (4.6)$$

From the equilibrium concentrations of the ions in the resin and the equilibrium concentrations in solution, at constant temperature, one can obtain the value of $a_{Na/H}$ or of $K_{Na/H}$. The plot, y_i vs x_i, is called ion exchange isotherm.

Figure 4.2 illustrates the plot of y_i vs x_i for a mono-monovalent equilibrium. The curve that gives a separation factor of 3, which gives a curve convex to the x-axis, represents a favorable equilibrium. The curve which is concave to the x-axis and gives a separation factor of 0.5 is an unfavorable equilibrium.

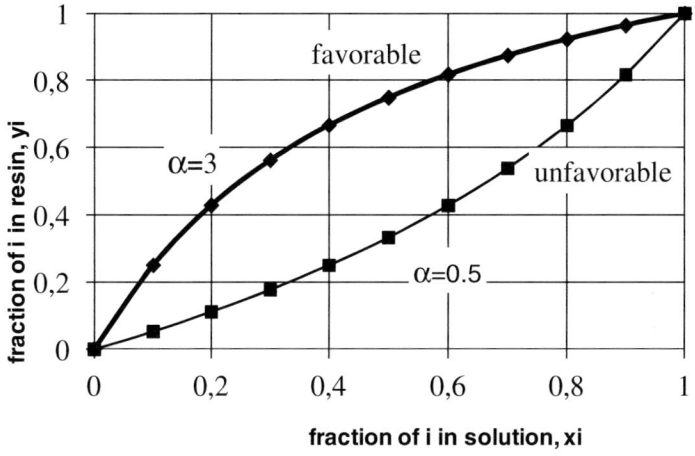

Figure 4.2 Isotherms for favorable and unfavorable equilibrium

For the general case:
$$a A + b R\text{-}B \leftrightarrows a R\text{-}A + bB \qquad (4.7)$$
we have:
$$K_{A/B} = \frac{q_A^a \, c_B^b}{q_B^b \, c_A^a} \qquad (4.8)$$

and

$$\alpha_{A/B} = \frac{q_A \, c_B}{q_B \, c_A} = \frac{y_A x_B}{y_B x_A} \qquad (4.9)$$

From (4.8) and (4.9) it results:

$$(a_{A/B})^b = K_{A/B} (q_A / c_A)^{b-a} \quad \text{or} \quad (a_{A/B})^a = K_{A/B} (q_B / c_B)^{b-a} \tag{4.10}$$

Therefore, the separation factor is not constant but it depends on the composition of the solution and the resin for the case where a is different from b. If we consider the case where ion A in eq. (4.8) is divalent and ion B is monovalent and therefore a=1 and b=2, and using fractions rather than concentrations (eq. (4.5)) then eq. (4.8) becomes:

$$\frac{y_A}{(1-y_A)^2} = K_{A/B} \frac{Q}{C} \frac{x_A}{(1-x_A)^2} \tag{4.11}$$

where Q is the total capacity of the resin in eq/L_R and C the total ionic concentration of the ions in the solution, in eq/L.

From eq (4.11) it is seen that for ions of different valence, the isotherm would depend on the total ionic concentration in solution, since the total resin capacity, Q, is constant. Figure 4.3 illustrates the case for a di-monovalent equilibrium using eq. (4.11).

It is interesting to observe that at low solution concentrations, the equilibrium is favorable while at high concentrations the equilibrium becomes unfavorable. Because of this feature, in a di-monovalent equilibrium, the divalent ion goes favorably to the resin during the loading cycle where concentrations are low but it goes favorably from the resin to solution during regeneration where concentrations are high.

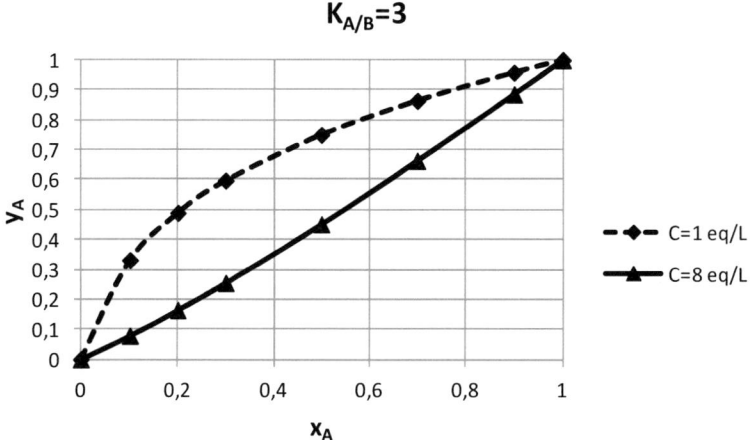

Figure 4.3 Ion exchange isotherms for a di-monovalent equilibrium with $K_{A^{2+}/B^+} = 3$. Curve with total solution concentration of 1 eq/L is favorable, curve with total solution concentration of 8 eq/L is unfavorable

Combining equations (4.10) and (4.11) one can calculate the separation factor $\alpha_{A/B}$ as a function of x_A. This is shown in figure 4.4 for the same value of K_{A^{2+}/B^+} as in figure 4.3. It is seen from fig. 4.4 that for C=8.0 eq/L where the equilibrium curve in figure 4.3 was unfavorable (concave to the x-axis), the value of $\alpha_{A/B}$ was in fact less than 1 while for C=1 eq/L it was higher than 1. It is also observed that, as it should be expected from the curves of figure 4.2, as the fraction of the divalent ion A, x_A, increases, the separation factor decreases for the favorable and increases for the unfavorable equilibrium.

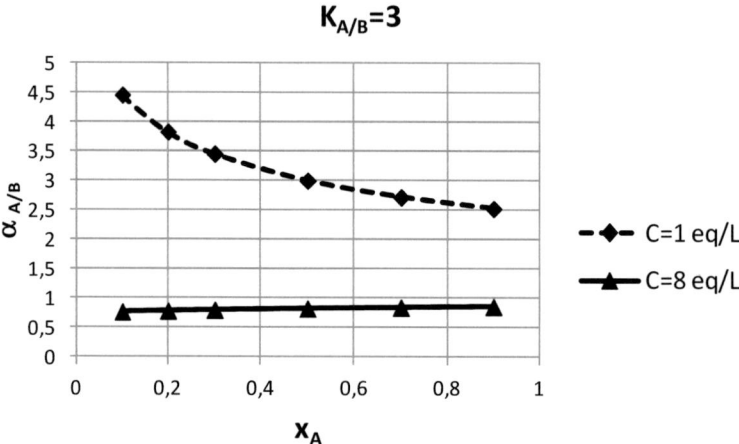

Figure 4.4 Separation factor $\alpha_{A/B}$ as a function of x_A

In the case of adsorption or on the removal of metals from solution with resins bearing ligands as functional groups, the reaction is:

$$R\text{-}L^{n-} + M^{n+} \leftrightarrows R\text{-}LM \quad \text{for the metals removal and}$$
$$A + S \leftrightarrows AS \quad \text{for adsorption}$$

where L is the ligand, M the metal, A the adsorbed species and S the free sites of the adsorbent where available for adsorption. An example is the formation of a complex of Zn^{2+} with an iminodiacetic type chelating resin:

The equilibrium constant for metal removal is:

$$K = (RLM)/(RL)(M) = q/(Q-q)c \qquad (4.12)$$

Where q is the concentration of the metal on the resin in moles/L_R, Q the total resin capacity in moles/L_R and c the metal concentration in solution at equilibrium, in moles/L.
A similar relationship holds for adsorption.

Equation (4.12) can be written:

$$q = KQc/(Kc+1) \qquad (4.13)$$

equation 4.13 is illustrated in figure 4.4 for different values of K, when Q=1.5 mol/L_R:

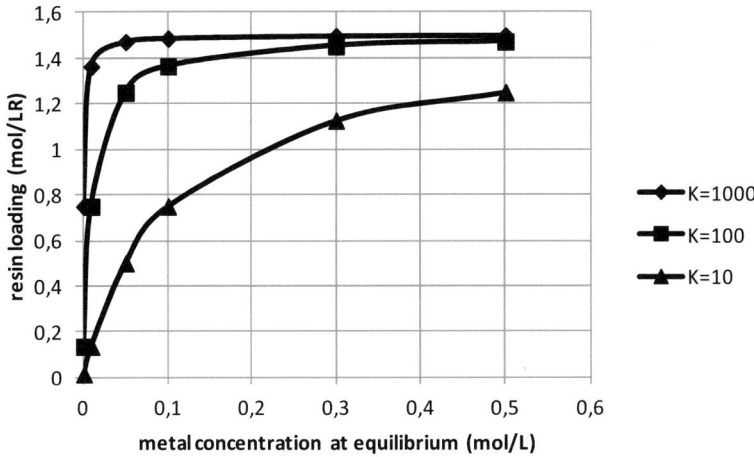

Figure 4.5 equilibrium isotherms in metal removal by complex formation.

Contrary to the binary equilibrium in ion exchange where when the feed solution contains only one of the components, the resin loading in a column at equilibrium is the total resin capacity (at $xi=1$, $yi=1$ see fig 4.1 or 4.2) no matter what the concentration of this component is, in metal removal or in adsorption, when the feed solution contains a metal or a compound, the equilibrium resin loading in a column depends on the metal or the compound concentration in solution, as figure 4.5 implies.

Selectivities

In multicomponent systems, we can establish a selectivity sequence by comparing the separation factors of various ions with respect to one ion taken as a reference. There are some rules that characterize the relative selectivities of a resin for different ions:

> *selectivity increases as the size of the (hydrated) ion decreases*
> *selectivity increases as the valence of the ions increases*
> *selectivity increases as the degree of crosslinking of the resin increases*

These rules can be explained qualitatively as follows.
Consider the sequence of the ionic radii of the hydrated cations from page 40:

$$H^+ > Li^+ > Na^+ > K^+ > Rb^+ > Cs^+$$

Due to swelling pressure the resin replaces easier a large ion by another one with smaller size. Consequently, the selectivity sequence of SAC exchangers for monovalent cations with respect to H^+ taken as a reference follows the reverse sequence of the hydrated radii of the ions (Helfferich, 1962):

$$Li^+ < Na^+ < K^+ < Rb^+ < Cs^+$$

This becomes more pronounced with resins with high degree of crosslinking. Again, with resins with very high degree of crosslinking where hydration of the ions may be incomplete, the selectivity can be reversed because the size sequence of the partially hydrated ions is reversed.

In the table below they are indicated as an illustration, the selectivities of SAC resins having different DVB content for various cations, based on selectivity coefficients (de Dardel and Arden, 1989):

TABLE 2.2

Cation	degree of crosslinking (% DVB)			
	4	8	12	16
H^+	1	1	1	1
Na^+	1.3	1.5	1.7	1.9
K^+	1.75	2.5	3.05	3.35
Zn^{2+}	2.6	2.7	2.8	3.0
Cu^{2+}	2.7	2.9	3.1	3.6
Ni^{2+}	2.85	3.0	3.1	3.25
Ca^{2+}	3.4	3.9	4.6	5.8
Pb^{2+}	5.4	7.5	10.1	14.5

Note that the selectivity coefficients for mono-monovalent equilibria are not directly comparable to those of, for example, di-monovalent equilibria. For example, a selectivity coefficient of a mono-monavalent equilibrium gives an equilibrium isotherm curve which may be the same, at a given total solution concentration, as a di-monovalent equilibrium having a lower selectivity coefficient (figure 4.6).

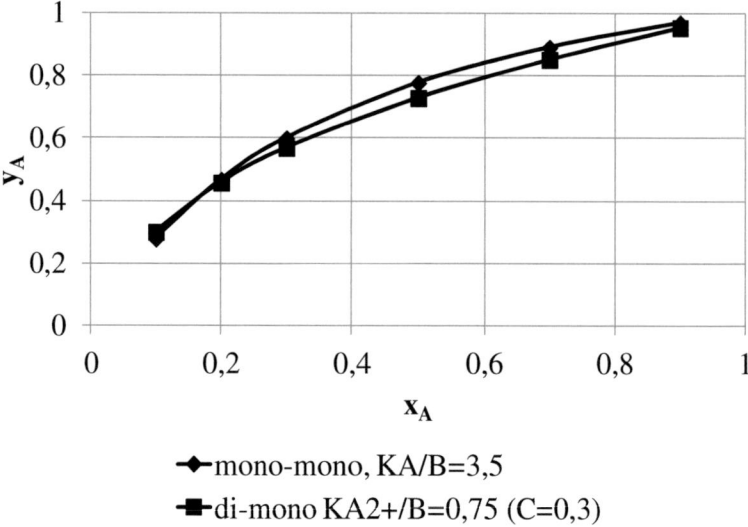

◆ mono-mono, KA/B=3,5
■ di-mono KA2+/B=0,75 (C=0,3)

Figure 4.6 Equilibrium isotherms for a mono-monovalent and a di-monovalent equilibrium. K is the selectivity coefficient, C is the total solution concentration.

In ion exchange, the counter-ions are located near the fixed functional groups and interact with each other by electrostatic attractions. The strength of this attraction is proportional to the ionic charge and inversely proportional to the square of the distance between the counter-ion and the fixed functional groups.

The attraction is higher when the valence of the ions is high and the distance between opposite charges is small.
Consequently, selectivity is influenced by the ionic radius of the hydrated ion of a given valence and by a given resin. The smaller the radius is, the closest will be the distance between counterions and fixed ionic groups and the higher will be the affinity of the resin for this ion.

Combining the effect of ionic charge and ionic radius, a SAC resin has the following sequence of selectivities taking the H^+ as reference (de Dardel and Arden, 1989) :

$$Ca^{2+} > Cu^{2+} > Zn^{2+} > Mg^{2+} > K^+ > NH_4^+ > Na^+ > H^+$$

As indicated previously, selectivity sequence based on separation factors for ions of different valence depend on the total ion concentration in solution.
Cation exchangers with the carboxylic functional groups show the opposite affinity series for alkali and alkaline earth metal ions compared to SAC resins (Gregor *et al*, 1956). This is due to the fact that the carboxylic group prefers larger, more hydrated and more polarizable counter-ions, the contrary of sulfonic resins. The affinity of carboxylic resins is therefore in the following sequence:

$$H^+ > Mg^{2+} > Ca^{2+} > Sr^{2+} > Ba^{2+} > Li^+ > Na^+ > K^+ > Rb^+ > Cs^+.$$

while for sulfonic resins is: $Li^+ < Na^+ < K^+ < Rb^+ < Cs^+$

Selectivities of carboxylic resins (Meyers, 1999):

Ion	Selectivity
Na^+	1
NH_4^+	2
Ca^{2+}	80
Cu^{2+}	160
Ni^{2+}	130
Pb^{2+}	>1000

In addition to the effect of the hydration of the counter-ions, selectivities are influenced by the degree of hydration of the resins. Resins with higher water content will prefer more hydrated ions. The water content of the resins depends on the crosslinking density and also on their chemical structure. For example, anion exchange resins exist with styrene-DVB and with acrylic-DVB matrix. Acrylic resins are more hydrophilic than styrenic due to the open chain aliphatic groups containing peptide links where water molecules can form H-bondings. As a result, styrenic resins prefer hydrophobic ions compared to acrylic resins. For example, the affinity of styrenic resins for chromates is higher than of acrylic resins (SenGupta and Clifford, 1986).

Although most of the cations are monoatomic spherical and we talk about ionic radii of hydrated ions, with anions it is not so simple because many of them are polyatomic and in these cases we talk about degree of hydration rather than ionic radius (SenGupta and Clifford, 1986). As mentioned before, resins with more hydrophilic matrix will have higher selectivities for the more hydrated ions and vice-versa, the more hydrophobic resins will have higher selectivities for the more hydrophobic ions. As an illustration, hydrophobic styrenic strong base anion exchange resins having long alkyl groups like tripropyl, tributyl etc have high affinity for hydrophobic anions like chlorates and perchlorates (Hofmeister series) based on their ability to precipitate a mixture of hen egg white proteins:

From strongly hydrated to weakly hydrated anions:

$Citrate^{3-} > sulfate^{2-} > phosphate^{2-} > F^- > Cl^- > ClO_3^- > Br^- > I^- > NO_3^- > ClO_4^-$

From weakly hydrated to strongly hydrated cations:

$N(CH_3)_4^+ > NH_4^+ > Cs^+ > Rb^+ > K^+ > Na^+ > H^+ > Ca^{2+} > Mg^{2+} > Al^{3+}$

As another example, from table 2.1, the radii of the hydrated halogen ions is about the same, while the non-hydrated radii increase from F^- to I^-. Consequently, the hydration of these ions should be in the following order:

$$F^- > Cl^- > Br^- > I^-$$

It is expected therefore that the selectivities of (hydrophobic) styrenic strong base anion exchange resins for these anions should be in the order (from less hydrated to more hydrated):

$$I^- > Br^- > Cl^- > F^-$$

This is indeed what has been found (Dow Technical Information "Using Ion Exchange Resins Selectivity Coefficients"), as shown in Table 2.3:

TABLE 2.3

Ion	SBA Type 1	SBA Type 2
I^-	175	17
Br^-	50	6
Cl^-	22	2.3
F^-	1.6	0.3
OH^-	1.0	1.0

In addition, although most of the commercial SAC are of styrene-DVB type with sulfonic functional groups, anion exchange resins exist with styrenic, acrylic or formophenolic matrix and the functional groups for the SBA can be of type 1 (trimethyl quaternary ammonium), type 2 (dimethylhydroxyethyl quaternary ammonium) or other groups such as tripropyl, tributyl or other types of quaternary ammonium and for the WBA can be a tertiary amine (polyvinyl benzyldimethylamine) or a tertiary amine made by amidation of dimethylaminopropylamine with polyacrylic acid. Phenolic WBA resins are made by polycondensation of phenol and formaldehyde functionalized with triethylenetetramine. These different structures greatly affect the affinities of different anions for these resins.

For a SBA resin we have, taking the OH^- as reference (Clifford, 1999):

$$[UO_2(CO_3)_3]^{4-} >> SO_4^{2-} > HSO_4^- > NO_3^- > Cl^- > HCO_3^- > OH^-$$

In a multicomponent system where more than two ions take part in the equilibrium, the above selectivity sequence holds provided that there are no interactions among the ions. In that case, given the values of $\alpha_{A/X}$ and $\alpha_{B/X}$, it follows that $\alpha_{A/B} = \alpha_{A/X} / \alpha_{B/X}$. If for example in a solution there are OH^-, Cl^- and NO_3^-

ions in contact with a SBA resin, the above selectivities hold provided that the presence for example of Cl^- does not interfere with the exchange of NO_3^- for OH^- or the presence of NO_3^- does not interfere with the exchange of Cl^- for OH^-.

Except the effect of hydration and size of the hydrated ions, some special interactions affect selectivities:

Ion pair formation. Ion pairing is a chemical interaction between the fixed functional group of the IER and the counter-ion. Examples are the strong affinity of sulfonic resins for Ba^{2+}.

Polarizability, whereby the distance between the ion and the fixed groups decreases. Examples: Ag^+ with SAC resins, metal cyanide complexes with SBA resins

Complex formation between the counter-ion in the resin and the ion in solution: example is the high selectivity of SBA in the SO_4^{2-} form for uranium due to the formation of the $UO_2(SO_4)_3^{4-}$ on the resin, as discussed in the following section.

The affinity of an ion for the resin increases due to the possibility of the ion to equilibrate on the exchanger phase in another form of higher valency, thus having higher affinity for the resin, for example the selectivity of SBA for chromate vs chlorides or sulfates is high because of the possibility of chromates to form dichromates on the resin phase which is fixed more strongly (SenGupta and Clifford, 1986)..

Steric configurations. Example is the divalent/monovalent selectivities of anion exchange resins affected by the distance of two adjacent functional groups (Clifford and Weber, 1982). Thus, the separation factor for SO_4/NO_3, $\alpha_{S/N}$, of phenolic resins is much higher than styrenic resins due to the favorable distance between two adjacent amine functional groups of these resins (see figure 2.12):

Resin	matrix	$\alpha_{S/N}$
Amberlite IRA400	sty/DVB	1.89
Amberlite IRA410	sty/DVB	2.40
Amberlite IRA458	acrylic/DVB	8.20
Duolite A7	phenolic	108
Duolite ES561	phenolic	109

Another case of steric effect on selectivity is the effect of the alkyl size of the alkylamine groups. When the alkyl group has more than one carbon atom, for example ethyl-, propyl- or butyl or even higher, it prevents polyvalent ions from approaching the nitrogen atom, thus making these resins selective for the monovalent ions. Thus, SBA resins with triethylamine groups are selective for NO_3^- over SO_4^{2-} ions, or tributylamine resins are more selective for the monovalent $Au(CN)_2^-$ than polyvalent cyanides of Cu, Zn, Ni or Co.

Possibly for the same reason, the selectivities for sulfates over chlorides of anion exchangers bearing amine functional groups are of the order (Clifford and Weber Jr, 1983):

Primary > secondary > tertiary > quaternary

Due to their high affinity for sulfates, WBA resins with primary and secondary amine functional groups have been suggested for the removal of sulfates from seawater or brackish waters (Boari, 1974).

Kinetics

The rate at which the exchange of ions reaches equilibrium is not instantaneous. It depends upon the mass transfer of the ion i in solution to the exchange sites in the resin and that of the ion j from the exchange sites into the solution. This mass transfer of ions from the solution to the resin and vice versa is essentially a diffusion process.

Figure 4.8 illustrates a resin bead in contact with a solution. Around the resin bead, there is a stagnant layer of liquid of a thickness δ, called Nernst film. Assuming that the transfer of ion Na^+ from the bulk of the solution to the Nernst film is instantaneous due to the agitation of the solution, the overall rate of ion exchange consists of the rate of the diffusion of Na^+ through the Nernst film and of the diffusion through the resin particle. If the rate of diffusion through the Nernst film is slow compared to the diffusion through the resin, the kinetics are film diffusion controlled. If on the contrary the diffusion through the resin particles is slow compared to the diffusion through the Nernst film, the kinetics are particle diffusion controlled.

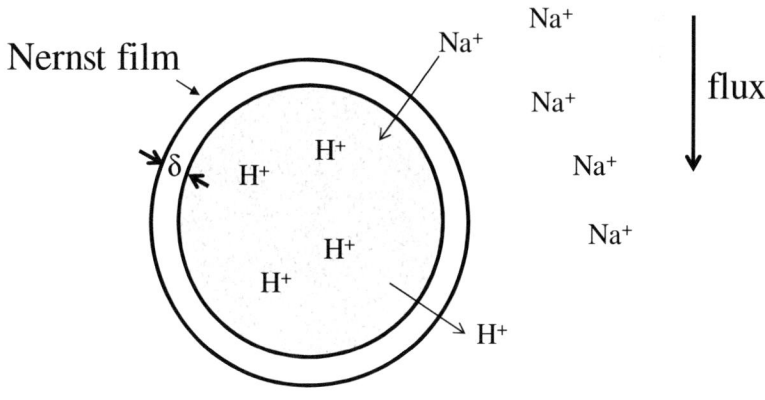

Figure 4.8

The diffusion of the Na^+ ions in the example of figure 4.8 into the resin particle is characterized by a diffusion coefficient, $\underline{D}_{Na+}$. This diffusion coefficient is slower than the diffusion coefficient D_{Na+} of Na^+ in solution. The reason is that in a resin particle, part of the space is occupied by the polymer matrix while the ions diffuse only through the water phase. There are other reasons as well, for example the path that ions follow is not a straight line but tortuous or the size of the ions in comparison to the size of the openings of the polymer network may be large. For these reasons, some models were developed to predict an effective diffusion coefficient $\underline{D}$ in relation with D. They give expressions varying between

$$\underline{D}=D(\varepsilon/2)$$

and

$$\underline{D}=D[\varepsilon/(2-\varepsilon)]^2$$

where ε is the fractional pore volume of the resin particle.

In order to derive the rate of ion exchange, that is the fractional attainment of equilibrium as a function of time, there exist two approaches. One is to set up the appropriate flux equations for the diffusing species and integrate them to obtain ion exchange rates. The other is to postulate rate equations involving reaction rates or mass transfer coefficients, which are easier to integrate.

Assume that the counter-ion in the resin particle is A and in the solution B and an exchange is taking place between A and B. Ion A is diffusing out of the resin due to its concentration difference between the resin and the solution. In order to preserve electroneutrality, an equal number of ions B (assuming equal valence) have to diffuse from the solution into the resin. If the mobilities of A and B are not the same, an electrical field will develop which tends to accelerate the slower ion and to slow down the faster one. We have therefore a flux where concentration difference is the driving force (and where Fick's 1st law applies)

$$J_{Ad} = -\underline{D}_A \, \partial \underline{C}_A / \partial x$$

and a flux where electrical potential gradient is the driving force

$$J_{Ael} = -\underline{D}_A \, \underline{C}_A (z_A F/RT) \, \partial \phi / \partial x$$

Where F is Faraday's constant, ϕ the electric potential, R the gas constant, T the absolute temperature.
The combined flux is given by the Nernst-Planck equation:

$$J_A = -\underline{D}_A \, [\partial \underline{C}_A / \partial x + z_A \underline{C}_A (F/RT) \, \partial \phi / \partial x]$$

From the flux equation one derives the concentration profiles

$$d\underline{C}_A / dt = -\text{div} J_A$$

and the reaction rate expressed as the fractional attainment of equilibrium, F(t) as a function of time.

In most cases, the solution of the differential flux equations involve great mathematical difficulties. For that reason, frequently it is prefered to take simplified differential reaction rate equations of the form

$$-dc_A/dt = f(c_A, c_B, q_A, q_B, \ldots)$$

which are more readily integrated. In fact, this approach was taken in the rate of loading tests described in the last chapter (page 237) and used in the design of the NIMCIX type columns (page 198).

The criterion on which it can be predicted whether the kinetics of ion exchange are particle or film diffusion controlled is (Helfferich, 1962):

$$\frac{X\underline{D}\delta}{CDr_0} (5 + 2\alpha_{B/A}) \ll 1 \qquad \text{particle diffusion control}$$

$$\frac{X\underline{D}\delta}{CDr_0} (5 + 2\alpha_{B/A}) \gg 1 \qquad \text{film diffusion control}$$

Where:
X= concentration of fixed ionic groups
C= concentration of solution (equivalents)
$\underline{D}$= interdiffusion coefficient in the ion exchanger
D= interdiffusion coefficient in the film
δ= film thickness
r_0= bead radius
$\alpha_{B/A}$ = separation factor = $\underline{C}_A C_B / \underline{C}_B C_A$ C in molarity

Factors that favor the film-controlled kinetics are those that accelerate the particle diffusion and vice versa:
- high resin capacity and high affinity for the in-coming ion; these factors tend to keep the ion concentration near the resin surface at low values
- high solution concentration

- low linear velocity of the solution through the column or inefficient agitation of the solution (film thickness increases)
- small bead diameter. In fact, both, film and particle diffusion controlled kinetics depend on the resin particle size but the kinetics in the film are proportional to $1/r_0$ while in the particle to $1/r_0^2$ (Helfferich, 1962).
- low degree of crosslinking accelerates particle diffusion

The rate of ion transfer inside the resin particle does not depend on the external conditions of concentration but depends on the particle size of the resin and the internal diffusion coefficients. The diffusion of ions through the resin beads depends on the structure of the resin. Highly crosslinked polymer phase result in low moisture, low swelling and slow diffusion.

Weak electrolyte, WAC and WBA resins, constitute a particular case since in their non-dissociated, H form for the WAC and free base form for the WBA resins, these resins are hydrophobic, they contain low moisture and diffusion is slow. As soon as they are converted to the dissociated form however, they swell considerably and diffusion increases accordingly. The particle controlled kinetics for these type of resins, follow a so called shrinking core model (Slater, 1991 ; Clifford, 1999) where the diffusion coefficient in the shell changes considerably going to the core of the resin bead (figure 4.9):

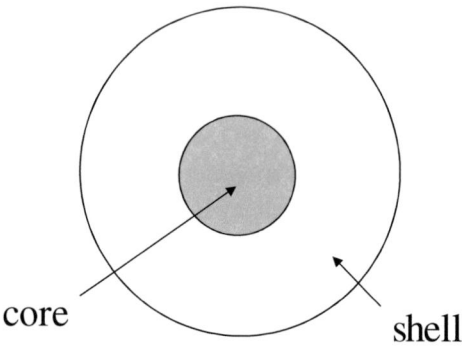

Figure 4.9

The above considerations apply to the gel type resins where the resin phase is homogeneous. In the case of porous resins, the resin phase is heterogeneous and contains macropores interconnected to each other and to the external bead surface in a way that form "channels" or "open pores". The kinetics of macroporous resins is therefore more complex. The macropores can be approximated with film kinetics. In this case, the mass transfer of the ions to the polymer phase depends also on the internal surface area of the porous resin beads. Therefore, for resins having the same particle size, the MR resins offer more surface area than gel resins. The access of the ion exchange sites then depends on the polymer phase in the microbeads : ionic access of MR resins is thus enhanced if the polymer phase in the microbeads is not too highly crosslinked.

5. Ion exchange processes

Ion exchange (IX) between resin and solution can be achieved by different techniques:
- batch operation
- column operation
  - single fixed bed
  - two or more columns in series
- continuous systems

Batch (equilibrium) operation

In a batch operation, the entire solution to be treated is mixed with an IER under agitation until after a certain time equilibrium is reached (fig. 5.1).
The extent to which the exchange of ions will take place depends on the selectivity coefficients or the separation factors for the given ions. Consider for example the exchange reaction (4.1) reproduced below where initially a SAC resin is in the H^+ form and the solution contains only NaCl.

$$R\text{-}H + Na^+Cl^- \leftrightarrows R\text{-}Na + H^+Cl^- \qquad (4.1)$$

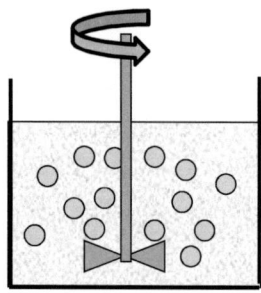

Figure 5.1 Batch operation

After mixing certain volumes of resin and solution and waiting for equilibrium, the resin will be found partially in the H^+ form and partially in the Na^+ form while the solution will contain a mixture of NaCl and HCl. For a given solution to resin volume ratio, the exact resin and solution composition will depend on the value of $K_{Na/H}$. The higher the value of $K_{Na/H}$ is, the more Na^+ ions will go onto the resin and the less will remain in solution. With only one batch contact however, the Na^+ ions will never go 100% onto the resin, some Na^+ concentration will always remain in solution.

For example, using equation (4.3) page 78:

$$K_{Na/H} = \frac{q_{Na}\, c_H}{q_H c_{Na}}$$

and the mass balance relationships:

$$C = c_H + c_{Na} \quad \text{and} \quad Q = q_H + q_{Na} \qquad (5.1)$$

one obtains:

$$K_{Na/H} = \frac{q_{Na}\, c_H}{(Q-q_{Na})(C-c_H)} \qquad (5.2)$$

where C is the initial solution concentration in eq/L and Q is the total exchange capacity of the resin in eq/L$_R$.

Using the fact that all H$^+$ ions in solution after equilibrium come from the resin initially 100% in H$^+$ form and the Na$^+$ ions found on the resin come from the solution initially containing only NaCl :

$$c_H = q_{Na} * V_R / V_S \qquad (5.3)$$

$$\text{or} \quad q_{Na} = c_H * V_S / V_R \qquad (5.4)$$

where V_R and Vs are the resin and the solution volumes respectively, one can solve equation (5.2) for c_H or q_{Na}.

Taking the value of 1.5 for $K_{Na/H}$, a ratio 5 for Vs / V$_R$, 0.1 eq/L initial concentration of Na$^+$ in the solution and a total resin capacity of 2 eq/L$_R$, we find that at equilibrium the Na$^+$ concentration in solution will be 0.015 eq/L or about 85% removal of Na$^+$ from the solution.

If we want to remove a high fraction of the Na$^+$ from the solution, a second batch contact has to be carried out with the same solution after equilibrium from the first contact and with a fresh resin in the H$^+$ form. After this second contact, the Na$^+$ ions remaining in solution will be less than after the first contact. These batch contacts can be continued for a number of additional contactors containing fresh resin, until the desired level of Na$^+$ in solution is reached. This is schematized in figure 5.2 Given the selectivity coefficient and the respective volumes of resin and solution, one can determine the number of stages of contact necessary to remove a certain fraction of Na^+ from the solution.

As an example, we can calculate how many batch contactors are necessary to remove 95% of the Na^+ from a 0.1 eq/L NaCl solution using a SAC resin with a total capacity of 2 eq/L_R (initially in the H^+ form), using a solution to resin volume ratio in each stage of 5:1. The selectivity coefficient $K_{Na/H}$ is 1.5

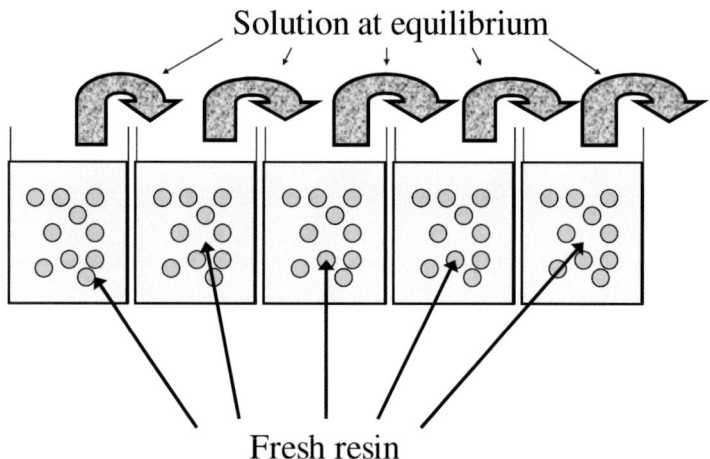

Figure 5.2 Multistage batch ion exchange

The calculations are analogous to those for one contact resin-solution except that in the second contact, the initial solution concentration has the values of H^+ and Na^+ calculated in the previous contact. This calculation is repeated a second, third time and so on until the desired value of the Na^+ fraction removed from the solution is obtained.

The results are illustrated in the figure 5.3 below. It is seen that under the conditions given above, two contacts are sufficient to remove more than 95% of Na^+ from the solution.

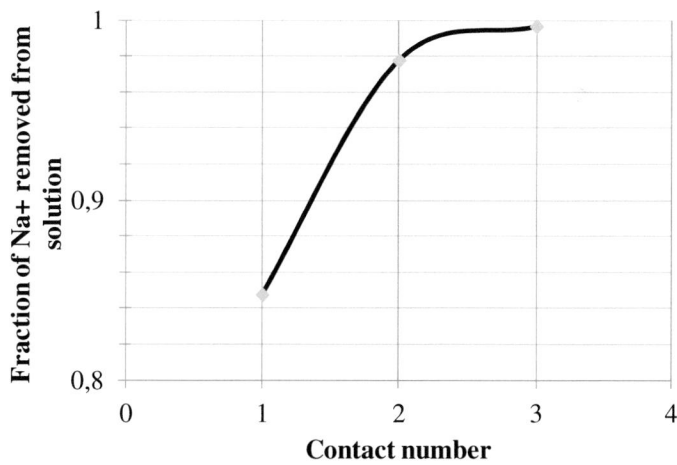

Figure 5.3

It is interesting to compare the above results with the case where we want to remove H^+ ions from a solution of HCl using a SAC resin in the Na^+ form. The exchange reaction will be the reverse of the reaction (4.1) page 78.

R- Na + $H^+ Cl^-$ ⇆ R- H + Na^+Cl^- (4.1a)

Again, the total exchange capacity of the resin is the same as above, 2 eq/L_R, and we use a solution to resin volume ratio in each stage of 5:1. The selectivity coefficient $K_{H/Na}$ is now

1/1.5=0.67, that is, the equilibrium here is unfavorable while before it was favorable. Following the same type of calculations as above, we obtain now the results shown in figure 5.4.

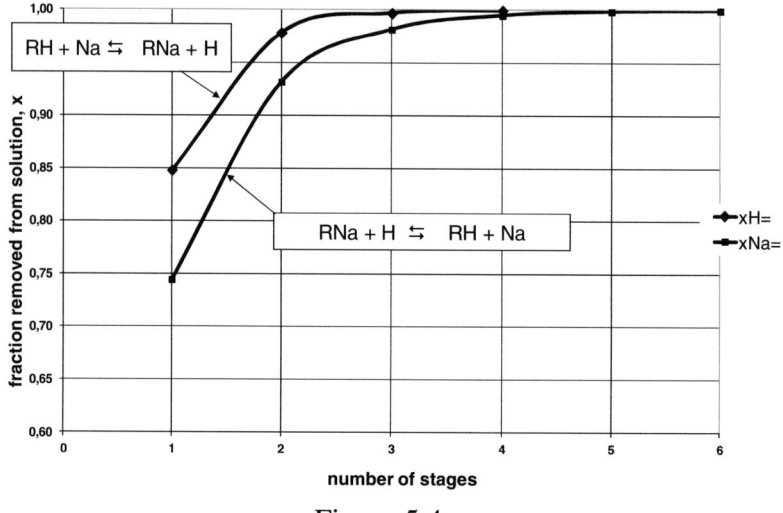

Figure 5.4

For comparison, the previous results of the case of Na^+ removal from the NaCl solution are also included. As seen, in the case of the unfavorable equilibrium, more stages are needed to remove the same fraction of the cation from the solution.

The situation will be different if we continue to pass fresh solution through the above batch containers, but leaving the resin as it was from the previous solution. As fresh NaCl solution enters the first compartment, equilibrium will take place with the already partially loaded resin, and so on with the resin in the rest of the compartments.

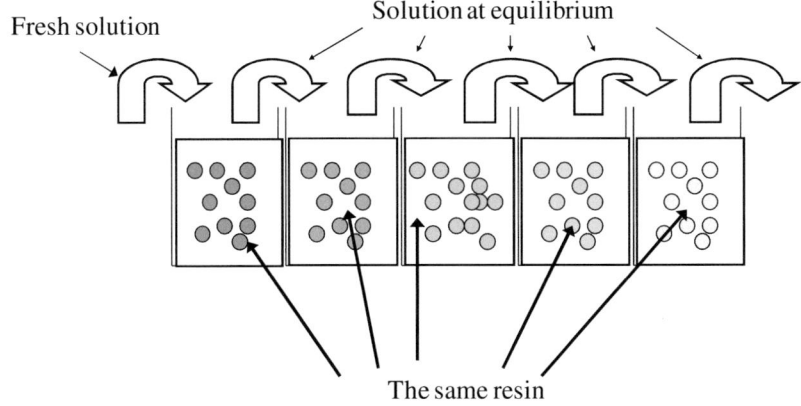

Figure 5.5 Dynamic batch ion exchange

After each new equilibrium, the resin in each compartment will contain more Na^+ ions and less H^+ ions. If this continues many times, a point will be reached that the resin in the first contactor will contain practically only Na^+ ions. We see in other words that if we continuously treat a fresh solution volume with a certain resin quantity, we can almost fully utilize the full resin capacity. This is illustrated in figure 5.5

Once again, one can calculate the resin and solution compositions at each contactor and after each contact resin-solution, using equations (5.1) to (5.4) and following the same procedure as above. In this way, it is possible to obtain the resin composition for various solution volumes that have equilibrated with resin in each of the contactors of figure 5.5.

Of these reactions, the equilibrium for (4.1) is favorable ($K_{Na/H}$ is 1.5) while that for the reverse reaction (4.1a) is unfavorable. We can calculate the fraction of H^+ remaining on the resin (for reaction 4.1) or the fraction of Na^+ remaining on the resin (for the reverse reaction 4.1a) as a function of the volume that passed through the contactors. The results are shown in figures 5.6 and 5.7 where we considered ten contactors in series. These figures show the resin composition, expressed as fraction of the resin in the H^+ or Na^+ form for reaction (4.1) and (4.1a) respectively, against the contactor number for different volumes (Bed Volumes, BV) passed through the contactors. Note that in these curves, by BV it is meant the contactor volume. A slightly different, but more generally used, definition of the Bed Volume is given in the next section.

As it is observed, the concentration profiles of the resin for the case of unfavorable or favorable equilibrium are completely different. This is discussed in more detail in the next section (Column operation).

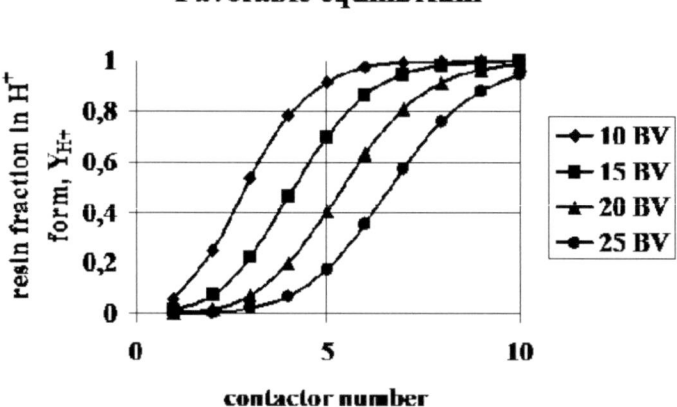

Figure 5.6 Resin concentration profile, favorable equilibrium

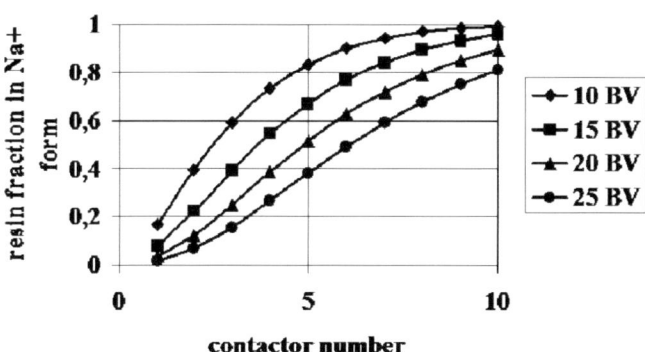

Figure 5.7 Resin concentration profile, unfavorable equilibrium

After the resin in the first contactor is saturated with Na^+ ions, we can remove this contactor from the circuit, pass a strong acid such as HCl through the resin, thus converting the resin back to its initial H^+ form, and then place this contactor at the end of the series, as last contactor. The conversion of the resin back to its initial form is called regeneration. In the recovery of uranium by IER, this step is called elution. Regeneration in uranium circles entails the treatment of the IER with NaOH to remove the silica that sometimes fouls the resins as it will be discussed more fully in Part II.

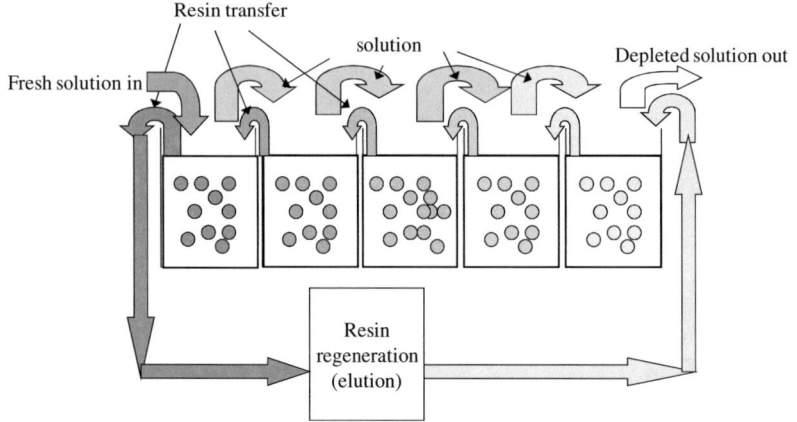

Figure 5.8 Continuous ion exchange

The process of periodically removing resin from the first contactor, regenerating it and then placing it as last contactor actually constitutes a continuous ion exchange (CIX) system. This is illustrated in figure 5.8.

In the examples of batch contactors in series of figures 5.2 and 5.5, with a given solution to resin ratio, the equilibrium constant, the concentration of NaCl in the feed solution and the total capacity of the resin it was possible to calculate at every contactor the equilibrium concentrations of Na^+ in solution and on the resin at every contactor, illustrated in figures 5.3 to 5.7. It is therefore also possible in the case of continuous ion exchange described here to calculate the equilibrium concentrations of Na^+, in each contactor for every batch of NaCl feed solution that enters the first contactor and progresses through the rest of the contactors to the end. In reality however, the feed solution does not enter the first contactor as a batch to remain in this contactor

until equilibrium, but it flows continuously through the contactors (fig. 5.8).
Provided that the solution flow rate is not excessive, it can be assumed that before the solution exits the contactors, equilibrium has been reached. Alternatively, it can be assumed that equilibrium concentrations have been only partially reached and the calculations are adapted accordingly. Typically, the residence time of the solution in a single contactor is about 15-20 minutes which is relatively long and the flow rate is slow.

Column operation

Single fixed bed column

In column operations, a solution containing ions to be removed flows through a column containing the resin in a certain ionic form.
In a single fixed bed column operation the solution flows continuously through a fixed resin bed, usually down flow. One can visualize the resin being a series of layers, or plates, with which the solution comes in contact, in a similar way as with a stirred tank in a batch operation. Depending on the kinetics of the ion exchange, after a certain contact time, an approximate equilibrium is reached in each layer. After contacting the first IER bed layer, the solution with the remaining ions from the first contact goes on further to the second layer, where it again moves towards equilibrium, resulting in an even lower concentration of the ions remaining in solution. As the solution progresses down through the resin in the column, the successive contacts with

layers of the resin further reduce ion concentration in solution until it attains a constant value. As fresh solution comes into the resin column, similar contacts take place with partially loaded resin and more of the incoming ions are removed. After a certain volume of feed solution has passed through the column, the upper part of the resin bed will become exhausted (loaded) and its composition will remain constant. The resin zone in which the concentration of the ion to be removed in the solution drops from the feed value to the final value is called ion exchange zone (IEZ) or reaction zone or front etc and is illustrated in figure 5.9.

Consider a binary ion exchange. If the equilibrium is favorable, and depending on the value of the separation factor, a reasonably sharp IEZ will be formed. As the solution moves down the resin bed in the column, a steady state is reached where the IEZ moves down without changing length (Fig.5.9a). This is called a "self-sharpening front". When a pre-determined ion concentration is found at the column outlet, the loading cycle usually stops (Fig.5.9b). This is termed the end-point, or breakthrough point, or operating capacity point. If the equilibrium is unfavorable, the ion concentration in solution diffuses as it goes down the resin bed and it never reaches a steady state (Fig. 5.9c). This is called a "broadening front".

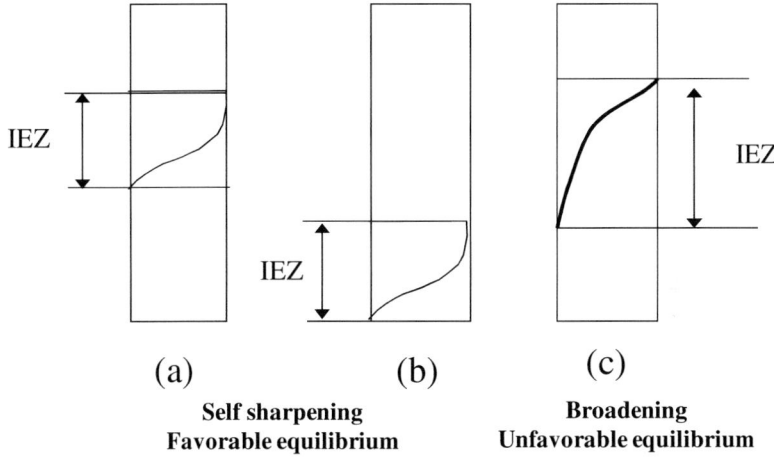

Figure 5.9 Ion exchange zone

These solution concentration profiles for favorable equilibrium showing a self-sharpening front and unfavorable equilibriul showing a broadening front are also illustrated in figure 5.10 for three different solution volumes that have passed through the resin bed.

It should be noted here that the formation of these fronts assumes a uniform resin bed free from imperfections, preferential channeling, foreign matter in the resin bed, air pockets etc, so that the solution flows through like a piston. This will be again discussed in the last Chapter, in *Column tests*, page 228.

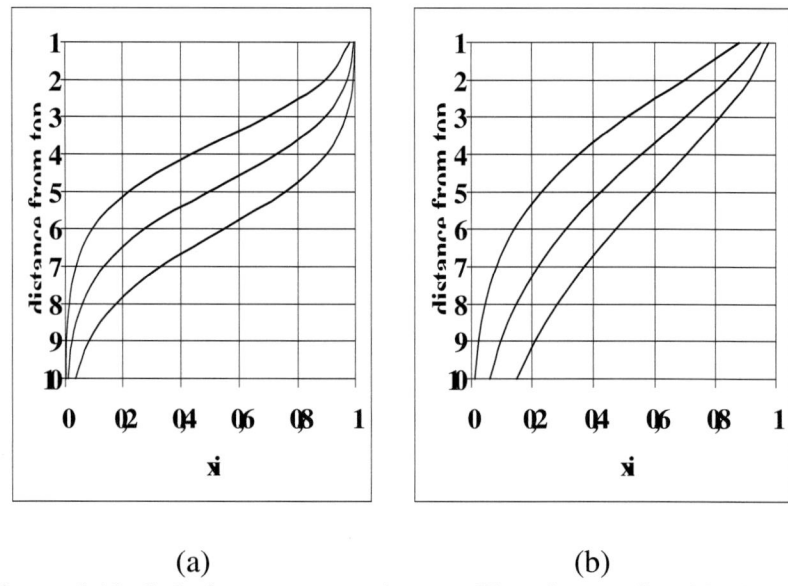

(a) (b)

Figure 5.10 Solution concentration profiles along a fixed bed ion exchange resin column for self-sharpening (a) and broadening (b) front for three different solution volumes.

Loading cycle

Another way of presenting column operation is by plotting the concentration of the ions exiting the column against the volume passed through the column, or versus time. Provided that the liquid flow rate is constant, these two ways are equivalent. In order to normalize the volume, Bed Volumes (BV) are used where:

$$BV = \text{(Volume of solution)} / \text{(Volume of resin)}$$

Similarly, for flow rate, bed volumes per hour, BV/h, is used. BV/h is called specific flow rate, as opposed to the volumetric flow rate, volume of solution per unit of time. The contact time of solution with resin is inversely proportional to the specific flow rate. Recalling that the linear velocity of the liquid is equal to the volumetric flow rate divided by the column cross sectional area, one can easily derive that the linear velocity is equal to the specific flow rate multiplied by the resin bed height.

Let us consider the following exchange reaction as an example:

$$R\text{-}H + Na^+Cl^- \leftrightarrows R\text{-}Na + H^+Cl^- \qquad (5.5)$$

The resin in the column is initially 100% in the H^+ form. As the solution of NaCl passes through the IER bed in the column, the Na^+ exchanges with H^+ and an IEZ will be formed. Below the IEZ, the resin remains 100% in the H^+ form while above the IEZ the resin has been converted 100% in the Na^+ form. At the beginning of the loading cycle, only H^+ and Cl^- ions are found in the solution coming out of the column, all Na^+ are retained by the resin. When the IEZ reaches the bottom of the column, the first Na^+ ions appear in the effluent solution together with H^+, and Cl^- ions. In fact the sum of the concentrations of $Na^+ + H^+$, in eq/L, is equal to the NaCl concentration in the feed solution. As the cycle continues, more and more Na^+ will appear until the Na^+ concentration in the effluent reaches the Na^+ concentration in the feed solution. The plot of Na^+ leakage vs volume (or time) is called "breakthrough curve".

It is possible to simulate these breakthrough curves by using various models (Slater, 1991). One model assumes that the column of the resin consists of a series of stirred tanks and at each tank the resin comes into equilibrium with the solution. Following are some examples of breakthrough curves obtained using the stirred tank model and using separation factors obtained experimentally. The number of tanks, or plates, was chosen to enable the breakthrough curves to closely simulate an experimental run.

Figure 5.11 illustrates the breakthrough curves for the ion exchange reaction (5.5).

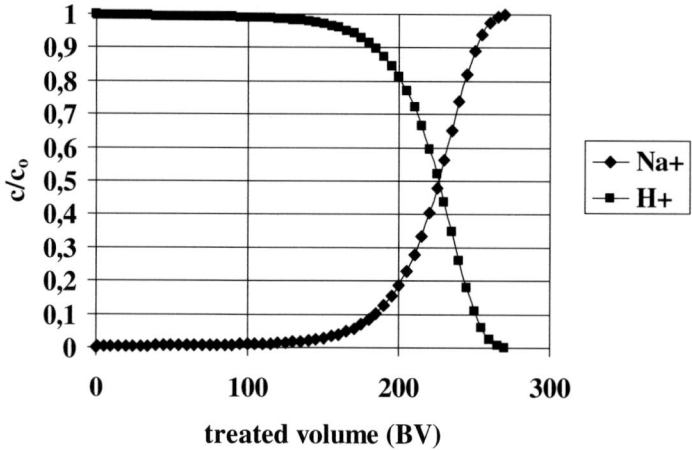

Figure 5.11 Breakthrough curves for the reaction
R- H + Na$^+$ Cl $\leftrightarrows$ R- Na + H$^+$ Cl$^-$

In multicomponent systems it is generally assumed that the exchange of each component does not interfere with the exchange of the other components. Now consider the case where KCl is added to the previous feed solution which now contains NaCl and KCl and the resin initially is 100% in the H$^+$ form. The selectivity coefficient $K_{K/H}$ is higher than $K_{Na/H}$. Initially, both Na$^+$ and K$^+$ will be exchanged for H$^+$ on the resin and H$^+$ will be displaced into the solution. From the selectivity coefficient we see that the resin prefers K^+ from Na^+. As a consequence, the K^+ will displace some Na^+ with the result that Na^+ will be predominant in solution. Two IEZ will be formed, one with K^+ / Na^+ followed by one with Na^+ / H^+. The resin above the first IEZ (that of K^+ / Na^+) will be partially in the Na^+ and partially in the K^+ form with the K^+ the predominant ion. Below the Na^+ / H^+ zone, the resin will be in the H^+ form. When the Na^+/ H^+ IEZ reaches the bottom of the column, Na^+ will start leaking and will eventually reach a maximum due to the displacement of Na^+ by the K^+ ions, before dropping to the feed concentration levels (figure 5.12).

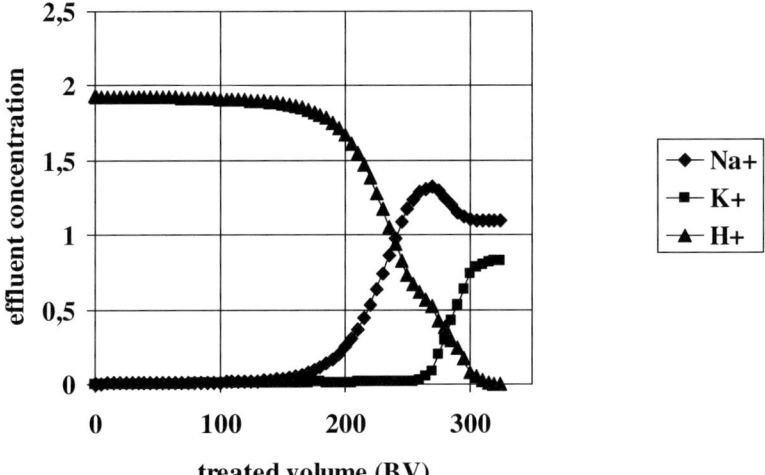

Figure 5.12 Breakthrough curves for the reaction
$R\text{-}H + Na^+Cl^- + K^+Cl^- \leftrightarrows R\text{-}Na + R\text{-}K + H^+Cl^-$

At the point where the K^+ / Na^+ zone reaches the bottom of the column, K^+ ions will start to leak. At saturation, the resin will be in a mixed K^+ and Na^+ form, the exact composition of which depending on the solution composition and the value of the separation factors.

Figure 5.13 illustrates additional breakthrough curves, again obtained by leakage simulations, for a water containing a mixture of HCO_3^-, NO_3^-, SO_4^{2-} and Cl^- ions (all as Na salts) and using the Imac® HP555 resin of Rohm and Haas Co. This resin is offered as a nitrate selective resin to remove nitrate from potable water. Note that the separation factor $\alpha_{NO3/Cl}$ of this resin is greater than $\alpha_{SO4/Cl}$, contrary to the general rule which states that selectivity increases as the valence of the ions increases. The separation factor $\alpha_{HCO3/Cl}$ is less than 1 and therefore the equilibrium in the exchange of HCO_3 for Cl is unfavorable (broadening front), contrary to $\alpha_{NO3/Cl}$ and $\alpha_{SO4/Cl}$ separation factors which are greater than 1 and therefore the corresponding equilibria are favorable (self-sharpening front).

This is seen in figure 5.13 where the shape of the breakthrough curve for HCO_3 leakage is convex to the x axis, while that of the NO_3^- and SO_4^{2-} leakage which are S-shaped. These breakthrough curves are in very good agreement with experimental data (Tebibel and Zaganiaris, 1992).

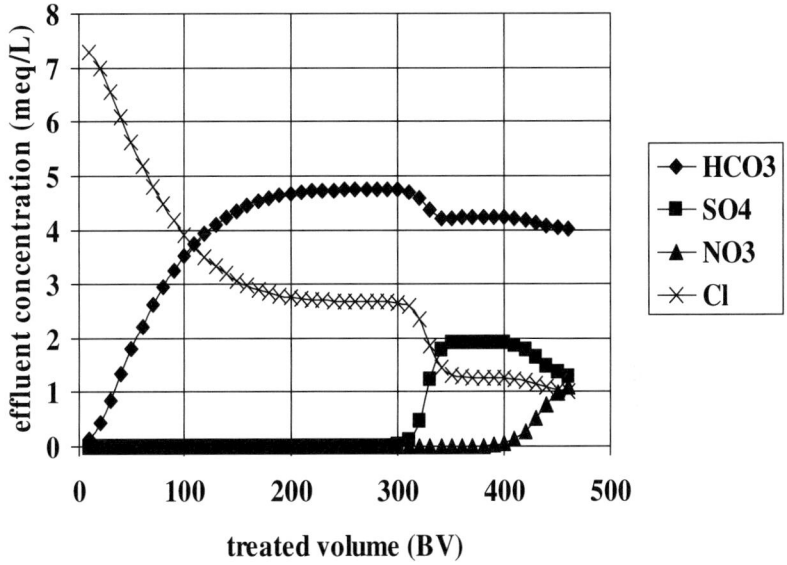

Figure 5.13

The other feature here is that the feed water also contains the Cl⁻ ion and the resin in this case is in the Cl⁻ form. As a result, the resin composition at saturation will contain the four ions, HCO_3^-, NO_3^-, SO_4^{2-} and Cl⁻ in varying proportions depending on the separation factors and the relative concentrations of the ions in the solution.

Figure 5.14 illustrates the resin composition close to saturation and this corresponds to the breakthrough curves of figure 5.13. Thus, at equilibrium, or at resin saturation, the resin will be found to be 5.7% in the Cl form, 6.9% in the HCO_3 form, 32.7% in the SO_4 form and 54.7% in the NO_3 form. Here we conclude that the resin composition at saturation in a column operation depends on the water composition and not on the initial resin composition.

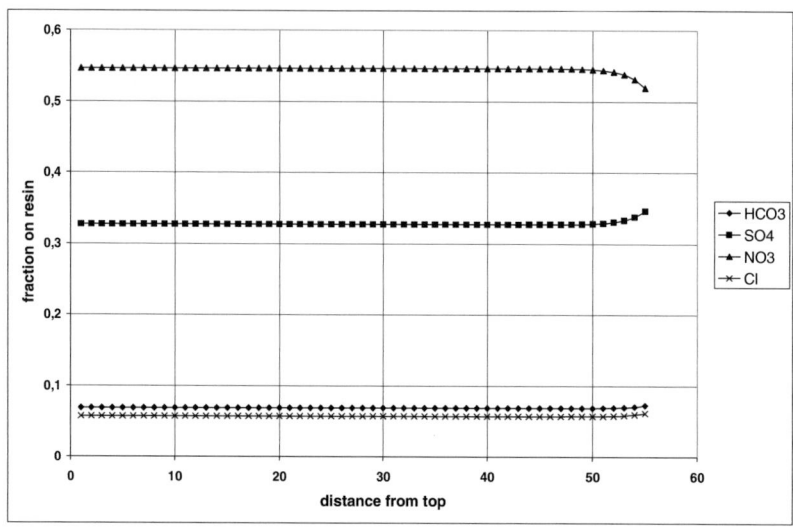

Figure 5.14 Resin composition at the end of the loading cycle with a water containing HCO3-, Cl-, SO42- and NO3-.

The IEZ of a self-sharpening front, illustrated in fig. 5.9a and b, corresponds to the distance from the end-point to saturation in a breakthrough curve, for example from 100 BV to 270 BV for the Na^+ leakage curve in figure 5.11.

In the hypothetical case where the selectivity coefficient is very high, the liquid flows through the resin bed like a piston and the ion exchange takes place instantaneously, the length of the IEZ will be extremely short and the IEZ in figure 5.9 becomes extremely sharp. In practice however, the imperfections of the resin bed caused by preferential channeling, axial dispersion of the ions or imperfections in the resin beads structure result in an increase in the length of the IEZ. In addition the selectivity coefficient and the kinetics of ion exchange cause further IEZ length increase. Factors such as flow rate and resin particle size that affect the extent of the equilibrium reached in each plate will affect the length of the IEZ. This is illustrated in figure 5.15 which shows two breakthrough curves, one where the kinetics of exchange is slow and/or the selectivity coefficient is low (curve I) and the other where kinetics are fast and/or the selectivity coefficient is high (curve II).

As shown in figure 5.15, when the leakage reaches a certain value, the breakthrough point, then the loading cycle stops and this determines the operating capacity of the resin.

The operating capacity (op cap) can be calculated using the expression:

Operating capacity = (ions fixed - ions leaked)*treated BV

For curve I of fig 5.15 the operating capacity is:

op cap = 0B * (feed conc.) − (surface of ABC)

Similarly, the operating capacity for curve II is:

op cap = 0E * (feed conc.) − (surface of DEF)

That is, curve II results in a higher operating capacity in comparison to curve I. Thus, a resin showing the characteristics of curve II will produce a longer service cycle and lower leakage. Often it is advantageous to have such a resin or to vary the operating conditions such that breakthrough curves similar to that of curve II are obtained.

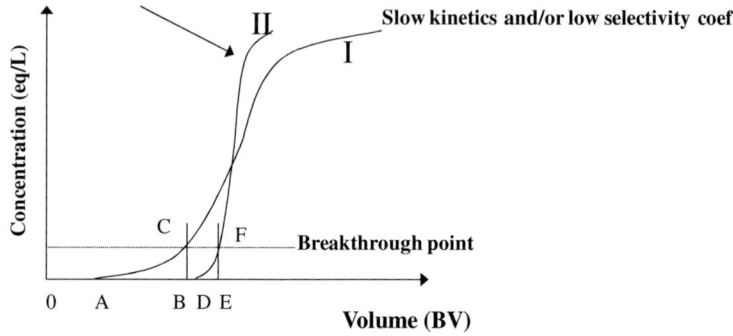

Figure 5.15

The particle size of the resin has an important impact on the kinetics of ion exchange and therefore on the operating capacity. Small size beads present a large surface area through which the ions can enter into the beads and diffuse through the particles. In addition, the Nernst film is thinner with small beads and the distance that the ions have to cover inside the resin particle to reach an exchange site is shorter. Consequently, resins with smaller particle size show faster kinetics and have higher operating capacities.

In figure 5.16 the effect of particle size on the breakthrough curves is illustrated assuming particle diffusion controlled kinetics and a favorable equilibrium. Except for the resin bead diameter, all the other operating parameters were assumed to be the same. As seen in fig. 5.16 the three curves cross each other at about the middle. In each case the saturation capacity is the same, the operating capacity to a given end point however varies, with the highest operating capacity obtained with the smallest particle size resin.

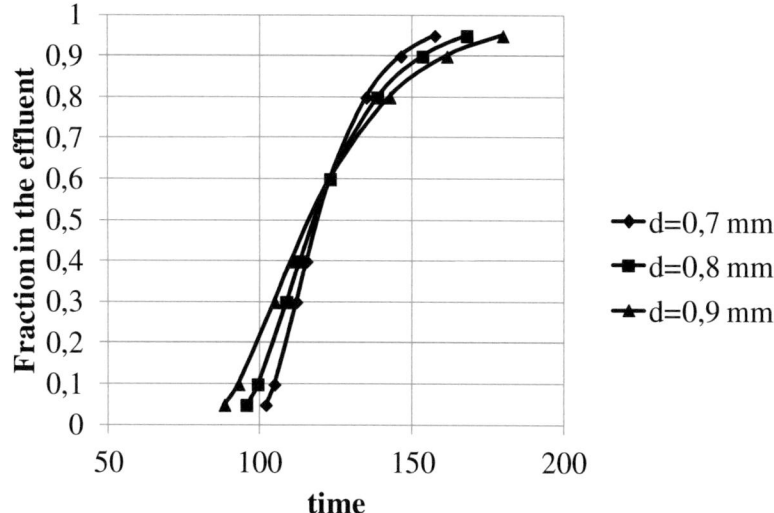

Figure 5.16 Effect of particle size on breakthrough curves (leakage simulations)

In a similar way, the breakthrough curve is also affected by the flow rate. In particle diffusion controlled kinetics, by increasing the volumetric flow rate of the solution through a given volume of resin, in other words by increasing the specific flow rate (BV/h), the incoming ions remain in contact within a layer of the resin bed for a shorter period of time, and thus move to the next stage prior to equilibrium being reached. The IEZ therefore becomes longer. Parameters that affect the diffusion of the ions through the particles include the resin particle size, the degree of crosslinking of the resin, the degree of conversion of the resin to the exhausted form and the size and the valence of the ions. Furthermore, when we consider film-controlled kinetics, where the rate of diffusion through the film is slower than that inside the resin particle, the linear velocity of the solution (the volumetric flow rate divided by the surface area of the empty column) affects the thickness of the film. At higher linear velocities the film becomes thinner and the ions have shorter path to travel to the particle surface. Therefore, by increasing the flow rate in this case, the length of the IEZ may increase but to a small extent.

As suggested in figure 5.9 (b), the higher the resin bed height in relation to the length of the IEZ, the larger the fraction of the ionic sites which will be used. Naturally, this implies that more resin will be needed to treat a given volumetric flow rate. What it should be avoided is to have a resin bed height shorter than the IEZ length as this will result in an unacceptably short cycle time. In cases where the IEZ length is high, like in the recovery of uranium, two or even three columns in series can be used. This is further discussed in the next paragraph.

In most cases, solutions which contain a high level of suspended solids and which are fed directly to a fixed bed of resin, will result in rapid plugging of the resin bed even to an extent that the solution flow through the bed could be curtailed. If complete suspended solids-liquid separation is not possible before ion exchange, the feed solution can be directed to flow upflow through the resin bed. In this case the resin, unless it is kept compacted by some physical means or air pressure, will fluidize. The degree of fluidization, as discussed in Chapter 3, will depend mainly on the linear velocity of the solution, the concentration of the suspended solids, the density and viscosity of the liquid and the temperature. In a fluidized bed operation the resin beads do not touch each other and thus the solution-resin contact will be less efficient than in fixed, compact beds. Consequently, for the same resin quantity, leakage during fluidized bed operation will be greater than in fixed bed operation.

Regeneration cycle

Ion exchange is a cyclic process. Because the ion exchange reaction is reversible, when the loading cycle ends, a regeneration, or elution step follows where the resin is stripped of its loaded ions and converted back to its initial ionic form. For example, if the loading cycle consists in removing Na^+ from a solution with a SAC resin in the H^+ form, eq (5.5), the regeneration consists in converting the resin back to the H^+ form using a regenerant solution such as HCl.

$$R\text{-}Na + H^+Cl^- \rightleftarrows R\text{-}H + Na^+Cl^- \qquad (5.6)$$

In an ion exchange reaction, if loading involves a favorable equilibrium then the regeneration (elution), which is the same reaction but in the opposite direction, should be an unfavorable equilibrium. This is illustrated as follows with a SBA resin. Take the situation of one SBA resin converted 100% into the Cl^- form with the eluent being NaOH and another situation where the SBA resin is converted 100% into the OH^- form where the eluent is NaCl. The ion exchange reaction is the same, only the direction differentiates the two cases:

$$R\text{-}Cl + Na^+OH^- \rightleftarrows R\text{-}OH + Na^+Cl^-$$

The separation factor for a gel type SBA resin, $\alpha_{Cl/OH}$, is about 15. This means that for the above reaction, the separation factor $\alpha_{OH/Cl}$ is $1/15=0.067$ and therefore the reaction is unfavorable. The elution profiles for the two cases are shown in figure 5.17.

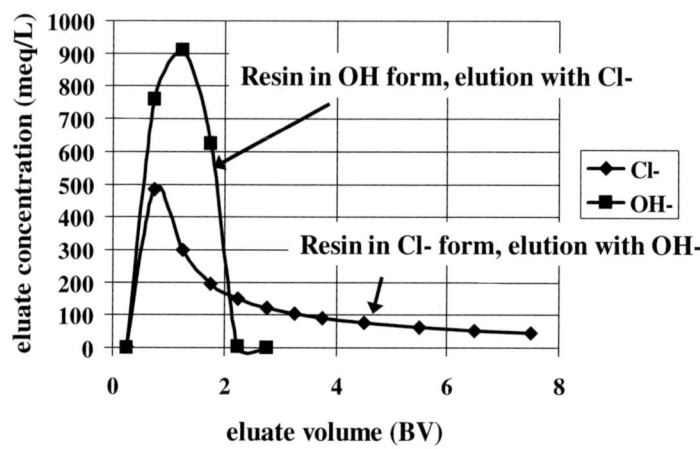

Figure 5.17 Elution profiles, flow rate 2 BV/h, eluent 1N NaOH or 1N NaCl, gel type SBA resin

As a result, the regeneration conditions, such as regenerant concentration and temperature, are chosen so that regeneration becomes as efficient as possible. In fact the quantity of regenerant employed is often a question of economics. The more regenerant used, the higher fraction of the resin will be converted to the initial ionic form. However, as it is seen from figure 5.17 above, most of the fixed ions are eluted during the first part of the regeneration while a large excess of regenerant is needed to remove the fixed ions down to very low level. Therefore, the regenerant level to be used is dictated by the cost and by the quality of the product solution during loading, in other words by the leakage tolerated.

One special condition which considerably affects the regeneration efficiency is the direction of the regenerant flow with respect to the feed solution during the loading step. As it can be seen in figure 5.9(b), with a down flow loading at the termination of the loading cycle the bottom part of the resin bed is only partially loaded while the upper part is saturated. When the regenerant flows down flow (co-flow or co-current regeneration, in the same direction as the feed flow), when the spent regenerant passes through the lower part of the resin bed, it saturates the partially loaded resin before it exits the column. Thus, the regenerant has not been fully utilized. Furthermore, in order to regenerate the lower part of the resin bed as completely as possible, a large excess of regenerant is necessary. If on the other hand the regenerant flows upward (counter-flow or counter-current regeneration, in the opposite direction to the feed flow), the lower part of the resin bed remains well regenerated whilst the upper part of the bed is only partially regenerated. The subsequent leakage in the following down flow loading cycle will be much lower in counter-flow regenerated beds than in co-flow regenerated beds.

All the above apply to strong electrolyre resins, that is, to SAC and to SBA exchange resins. For weak electrolyte resins, WAC and WBA exchange resins, since these resins at a slight excess of regenerant, acid for WAC resins or caustic for SBA resins, they are not ionized, they release all ions they have fixed and are fully converted to the regenerated form. Thus, the weak electrolyte resins are fully regenerated with a stoichiometric amount of regenerant, plus a small excess. The regeneration mode, co- or countercurrent, makes no difference, except if some compounds are fixed on these resin by a different mechanism, for example by adsorption.

In general, the regeneration step consist of :
- a backwash
- the regenerant injection
- a slow (or displacement) rinse
- a fast rinse

The backwash step is designed to remove any foreign particles or resin fragments and to thus clean and decompact the resin bed. This ensures that the in-service resin bed pressure drop is kept to the minimum. For an efficient backwash, the free space above the resin bed should be approximately 60% or more and this expansion should be maintained for 15-20 minutes or until the resin bed is clean.

The regenerant injection time should be sufficient for an efficient use of the regenerant. In the case of small ions with fast kinetics, this is in the range of 30-40 minutes for a standard particle size resin.

The slow rinse is designed to displace the regenerant still remaining in the resin bed and is carried out at the same flow rate as the regenerant flow rate.

The fast rinse is designed to remove the last traces of regenerant and is normally carried out at a flow rate of at least 80% of the normal service flow rate.

In calculating the resin requirements for an installation, the above considerations of operating capacity and its dependence on the specific flow rate and the linear fluid velocity, pressure drop and regeneration requirements should be taken into account. Following is a simple example of a calculation of an installation containing one fixed bed column of resin.

Consider a water softener with a SAC resin, which is the simplest case involving only one column of resin.

In softening, the objective is to remove hardness ions (Ca^{2+}, Mg^{2+}) and replace them with Na^+ ions, thus:

Loading:
$$2\text{ R-Na} + Ca^{2+} \leftrightarrows R_2\text{-Ca} + 2\text{ Na}^+$$

Regeneration:
$$R_2\text{-Ca} + 2\text{ Na}^+ \leftrightarrows 2\text{ R-Na} + Ca^{2+}$$

Take a water containing 10 meq/L Na^+ ions and 3 meq/L hardness ions at a flow rate of 100 m^3/h and using a 10% NaCl solution as a regenerant.

Initially, a SAC resin in the Na^+ form will be used. From information supplied by the resin manufacture, or from experimental trials, we find that for the above water composition, at a given specific flow rate, fluid linear velocity and a given regenerant level in co-flow or counter-flow direction, an operating capacity of 1 eq/L_R, or 1000 meq/ L_R, can be expected. The average hardness leakage will be 0.2 meq/L. If 1 liter of resin can remove 1000 meq of hardness ions (1 eq/ L_R), then since 1 liter of water contains 3 meq of hardness, one liter of resin will be able to treat 1000 / 3 = 333 liters of water, in other words, a given volume of resin will be able to treat 333 BV of feed water.

Next, one needs to define if he wants to dimension the installation based on cycle time or based on specific flow rate. If we decide to have a loading cycle of 12 hours, with the water flow rate of 100 m^3/h, a total of 12*100 = 1200 m^3 of water will be treated. Since this volume represents 333 BV, it follows that the BV, or the resin volume, will be 1200/333 = 3.6 m^3. This will make a specific flow rate of 100/3.6 = 27.8 BV/h. this specific flow rate should be compatible with the assumed operating capacity. From the resin data sheets we see that the required service water volume needed for backwashing, regenerant dilution and rinse, is about 15 BV or 54 m^3. The total water to be treated should therefore be 1200+54=1254 m^3 and consequently, the resin volume needed would be 1254/333= 3.76 m^3.

The dimensions of the column, given the resin volume already calculated, is approximately chosen so that the ratio resin height to diameter (H/D) to be not far from 1, for example between 0.8 and 1.2. Choosing a column diameter of 1.6 m (1600 mm) the crossectional area will be 2 m^2 the resin bed height will be 3.76 / 2 = 1.88 m with H/D = 1.2. The linear velocity during the loading cycle will be 100 m3/h by 2 m2 column cross-section, or 50 m/h.

The 50 m/h fluid velocity is considered too high from the pressure drop point of view. The calculation is therefore revised to obtain a lower fluid velocity and eventually a lower specific flow rate. For example, a 1.8 m diameter column would give a fluid velocity of 39 m/h and a resin height of 1.5 m (H/D = 0.8).

At 20 BV/h specific flow rate we obtain 5 m3 of resin, column diameter 1.8 m, resin height 2 m (H/D = 0.8), linear velocity 39 m/h and. the cycle time becomes 333/20 ≈ 17 hours.

With a cycle of 24 hours, the same considerations give a resin volume of 7.5 m3, column diameter 2 m and a linear velocity of 31 m/h.

If a continuous production of treated solution is desired, then there should be two columns in parallel, one on loading while the other one will be on regeneration or standby. When the column on loading reaches the end-point, the solution flow switches to the other column while the exhausted column is regenerated. Regeneration would be with 10% NaCl, or 100 g NaCl/L which means that in order to obtain a level of 150 g NaCl/L_R we will need 150 / 100 = 1.5 BV of 10% NaCl per regeneration. More detailed designs for water demineralization are discussed elsewhere (Applebaum, 1968).

Three columns in series (merry-go-round) system

From figure 5.15 it is seen that if the loading cycle stops at a point where the resin is near saturation, then the operating capacity reaches a maximum corresponding to the equilibrium value of the feed solution concentration. However, by the time saturation occurs the leakage is very high and probably unacceptable by the process parameters. If we stop the cycle at the breakthrough point and if the IEZ is very long, then only a small fraction of the total resin capacity is utilized during the loading cycle. In order to have both, good resin utilization and low leakage, a second column, with the same resin volume as the first column, can be added in series. The ions leaking from the first column will be taken up by the second column which in turn acts as a polisher, to give the lowest on-line leakage condition. When breakthrough is evident at the outlet of the second column, the cycle stops for a short period of time and the first column goes to regeneration, the second column becomes the primary column and a third, freshly regenerated, column goes to the polishing position. The loading cycle then continues. This system is called merry-go-round system (fig. 5.18). In a well designed merry-go-round system, when the tail column breaks through, the head column is close to saturation.

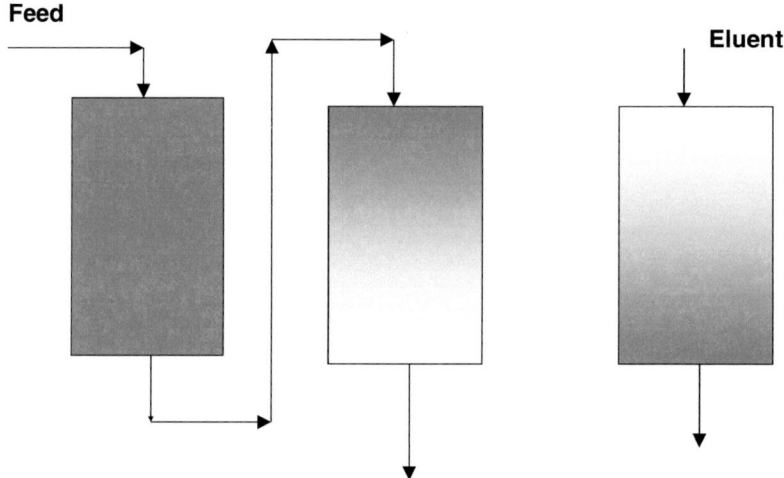

Figure 5.18 Three column merry-go-round system

The advantage of this system is
- in a well designed merry-go-round system, the resin inventory can be less than two single parallel columns.
- Since the head column that goes to elution is loaded close to saturation, the eluate contains a higher concentration of the eluted ions This results in a more efficient usage of eluent.

Design calculation for a three column merry-go-round system

Assume that we need to remove uranium from a carbonate leach solution containing 300 mg U_3O_8 /L using a three column merry-go-round system and at a flow rate of 100 m^3/h. Elution is with 3 BV of 1N $NaHCO_3$ in over 9 hours.

Since the kinetics of uranium recovery are slow, the operating capacity depends heavily on the specific flow rate. In a three column merry-go-round system, two columns are on loading and one on regeneration. If we choose to operate at 5 BV/h this applies to the total resin in loading (both columns) and therefore, for each column the specific flow rate will be 10 BV/h. The resin requirements are calculated from this specific flow rate as follows: for 100 m^3/h and 10 BV/h per column, the resin volume per column will then be 100 / 10 = 10 m^3. It should be noted that the choice of the specific flow rate (here: 5 BV/h) is made based on experimental trials with two columns in series. The system must run for several cycles until the resins in the two columns have reached a steady state condition. The appropriate specific flow rate is the one where when the second column breaks through, the first column is close to saturation.

In this system, operating capacity refers to the quantity of uranium fixed on the first column, which is almost the saturation capacity. The operating capacity should either be taken as the saturation capacity for the given leach solution, assuming the second column is at about the breakthrough point, or it should be determined experimentally with two columns running in series.

If we assume that the operating capacity is 40 g U_3O_8/L_R then, the loading cycle time will be (40 / 0.3)/10 = 13.3 h where 0.3 is uranium concentration in g U_3O_8/L and 10 is the specific flow rate per column. We conclude that this loading time is sufficient to allow for full elution of the third column.

If the resin capacity were for example only 20 g U_3O_8/L_R then the loading time would have been 6.7 h and this is insufficient for the required 9 hours needed for elution. In this case we need to either add more resin per column, for example (20/0.3)/9 = 7.4 BV/h or 100/7.4 = 13.5 m^3 resin per column to have a 9 hour loading cycle or we need to reduce the elution time to 6 hours. This will require an increase of the rate of elution and the overall eluent volume to ensure that the residual uranium on the eluted resin remains the same as before.

Continuous systems

The three column merry-go-round system is the simplest continuous system. Suppose we have, instead of two columns in series, six columns in series where when the last column breaks through, the first one is close to saturation while from the second to fifth, the columns are progressively less and less loaded. When the last column breaks through, the first goes to elution, the second column becomes head column and so on and in the end, a freshly eluted column is added. This configuration necessitates that the time of elution is at least the same or less than the time the last loading column goes to breakthrough. If this is not the case, then more columns need to be added to the elution section to enable fresh eluted columns to be available when the loading column breaks through. For example, if the elution time is three times that of the time to loading breakthrough, then three columns should be in the elution section. In this case, the eluent flows in counter-current direction of the columns that come to elution.

These principles are illustrated in figures 5.19 and 5.20. In figure 5.19, the absorption section contains six columns. The pregnant feed solution flows from column A1 to A6 and from column A6 it goes to the barrens. The shades of the columns in the figure indicate the degree of exhaustion/loading of the resin: the darker the color, the more exhausted the resin is.

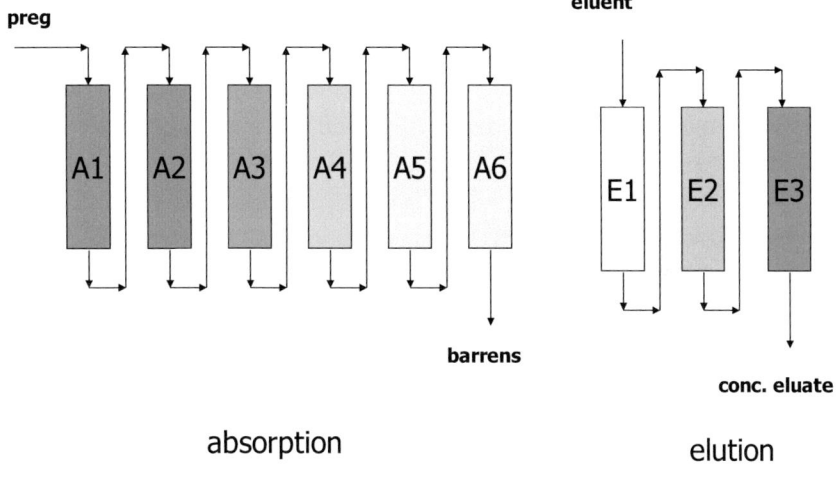

Figure 5.19 Continuous system, principle

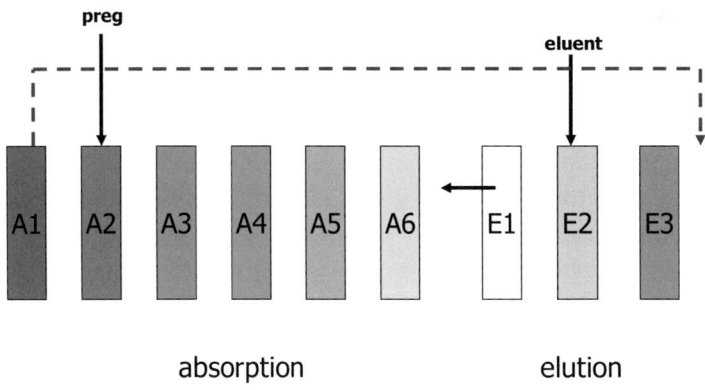

Figure 5.20 Continuous system, resin transfer

The elution column contains three columns and fresh eluent flows from column E1 to E3. From column E3 it goes to the concentrated eluate tank. Every a period of time that corresponds to the time it takes for breakthrough from column A6, say 4 hours, the pregnant solution flow and the eluent flow stop, column A1 is removed from the circuit and the pregnant is connected to column A2. Column E1 is rinsed and is connected to the exit of column A6 while column A1 takes the place of column E3 (fig.5.20) after it has been rinsed before to take feed solution out.

In practice, instead of having many individual absorption columns and many elution columns and moving around the columns, one can have one big absorption column divided in some way in segments (stages) each stage representing one of the individual small columns. Similar considerations apply for the elution column. In this case, instead of transferring the columns as described above, it is the resin segments that are transferred.
At the end of the loading cycle, flow of the pregnant solution and the eluent stop and resin from stage 1 of the adsorption column is transferred to a measuring chamber hydraulically. Resin from stage 2 is transferred to stage 1, from stage 3 to stage 2 and so on finally resin from the last stage to the next to last stage. Then resin from the first stage in the elution column which is the resin that has seen fresh eluent is transferred to the last stage in the absorption column. Then cycle starts again. Continuous systems used in uranium recovery are discussed in Chapter 9.

Part II. Uranium recovery

6. Uranium extraction

A stamp issued in 1965 by the Republic of Gabon showing the operations of the Mounana uranium mine. This mine was operated by COMUF, subsidiary of AREVA, from 1958 till 1999.

Uranium has four oxidation sates, III, IV, V and VI but only the IV and VI are stable. In the IV oxidation state Uraninite (UO_2) is the most common ore and it is slightly soluble. In oxidizing environment, U(IV) oxidizes to U(VI) whose compounds are soluble. In the presence of oxygen, U(IV) oxidizes and dissolves in water at low pH as the cation UO_2^{2+}:

$$2\ UO_{2(s)} + 4\ H^+_{(aq)} + O_{2(g)} \rightarrow 2\ UO_2^{2+}{}_{(aq)} + H_2O_{(l)}$$

UO_2^{2+} can form complexes with hydroxides, carbonates, sulfates and phosphates. Thus, at low U concentrations and up to pH of 5 in the absence of ligands, UO_2^{2+} is the predominant species. At higher pH, neutral and anionic hydroxyl complexes prevail (Rosenberg *et al*, 2016). In presence of carbonates in the water, up to a pH of 5, UO_2^{2+} is again the predominant species, above that, UO_2CO_3, $UO_2(CO_3)_2^{2-}$ and $UO_2(CO_3)_3^{4-}$ complexes become important (Riegel and Höll, 2008).

Crystal structure of UO_2. Canadian stamp issued in 1980 entitled Uranium Resources.

Uranium is extracted from the ore by open-pit or underground mining or by in-situ leaching (ISL). In mining from the ground, uranium extraction is based on the following steps: Crushing and grinding, leaching, solids-liquid separation, ion exchange (IX) and/or solvent extraction (SX) and yellow cake precipitation.

Uranium is extracted under agitation leaching or heap leaching. Depending on the ore composition, leaching is performed with sulfuric acid or with solium carbonate/bicarbonate. With autoclave leaching, the "pulp", that is the pregnant leach solution (PLS) and the fine ore particles, is either separated in liquid and the solids or treated as such directly with IX, a technology called Resin-in-Pulp (RIP) which eliminates the need for solids-liquid separation.

In the ISL technology, a suitable leach solution is injected into an ore zone. The leaching solution solubilizes uranium and then is pumped to the surface where uranium is recovered by IX. In general, the PLS in the ISL technology compared to conventional mining from the ground, contains a low uranium concentration and it has a different composition because of the almost closed loop circulation of the solutions without any removal of salts.

Figure 6.1. Heap leaching

Uranium forms strong complexes with phosphates, carbonates, fluorides, moderately strong complexes with sulfates and weak complexes with chlorides. Consequently, the recovery of uranium from acidic or alkaline leach liquors using ion exchange technology is based on the formation of carbonate or sulfate anionic complexes with the lixiviant, that can be fixed on an anion exchange resin. Other metals which do not form strong anionic complexes with the lixiviant, such as Fe^{2+}, Co^{2+}, Ni^{2+}, Cu^{2+} or Zn^{2+} are not fixed on an anion exchanger. It is also possible to recover uranium from acid leach liquors on a SAC resin as the cationic species UO_2^{2+} but these resins are not selective for UO_2^{2+} versus other cations and therefore this approach is not practically used (Zontov, 2006)

The selectivity sequences of strong base anion exchange resins for acid or alkaline leach are in the order (McGarvey and Ungar, 1981):

Acid leaching:
$V_2O_7^{4-} > Mo_8O_{26}^{4-} > UO_2(SO_4)_3^{4-} > UO_2(SO_4)_2^{2-} > Fe(OH)(SO_4)_2^{2-} > SO_4^{2-} > Fe(SO_4)_2^{-} > NO_3^{-} > HSO_4^{-} > Cl^{-}$

Alkaline leaching:
pH9-10: $\quad V_2O_7^{4-} > UO_2(CO_3)_3^{4-} > MoO_4^{2-} > UO_2(CO_3)_2^{2-} > SO_4^{2-} > CO_3^{2-} > NO_3^{-} > Cl^{-} > HCO_3^{-}$
pH 11: $\quad UO_2(CO_3)_3^{4-} > VO_4^{3-} > UO_2(CO_3)_2^{2-} > MoO_4^{2-} > SO_4^{2-} > CO_3^{2-} > NO_3^{-} > Cl^{-} > OH^{-}$

After loading, uranium is eluted from the anion exchange resins using a suitable eluent. From the concentrated eluate, uranium is recovered in relatively pure condition. The ion exchange operation therefore has essentially two functions, concentration and purification.

When chloride or nitrate elution is practiced, uranium is precipitated from the concentrated eluate using for example NH_4OH, where uranium precipitates as ammonium diuranate (ADU) or NaOH where it is recovered as sodium diuranate (SDU) at a pH of 12, or with H_2O_2 where uranium precipitates as uranium peroxide (UO_4).
If precipitation is done in two stages in order to eliminate first iron, lime is added to a pH of 3.7 where iron is precipitated, followed by further neutralization with NH4OH or NaOH to precipitate uranium.

If H_2SO_4 elution is practiced, uranium is recovered from the concentrated eluate using solvent extraction (SX). This process is called Eluex process (Bufflex in South Africa). Then uranium is stripped from the solvent using a number of stripping solutions such as $(NH_4)_2SO_4$ and NH_4OH or H_2SO_4. From the strip solution uranium is recovered by precipitation.
Prior to elution, a scrubbing step can be included where impurities such as iron are removed from the resin, leaving the uranium in purer condition, and at the same time increasing the uranium concentration in the resin thus resulting into a more concentrated eluate.
The purity of the recovered uranium depends on the impurities found in the eluates. One of these impurities in acid leaching is Fe^{3+} which is fixed on the SBA resins as a sulfate complex along with uranium and is eluted together with uranium. Other impurities that can be fixed on the SBA include chlorides, nitrates, phosphates, thorium, molybdenum and vanadium (Udayar *et al*, 2013).

The preparation of the feed solutions varies considerably for different ore types. Acid leaching using H_2SO_4 is a very frequently used process being of low cost and high efficiency. From low grade ores where the resulting acidic pregnant solution contains relatively low concentration uranium (below about 500 ppm), ion exchange is the preferred technology. For high grade ores resulting in uranium concentrations higher than 500 ppm, SX is practiced. SX is also preferred even at uranium concentrations below 500 ppm in cases where the pregnant solution composition is such that low operating capacities are obtained with IER, for example in presence of very high chloride or iron concentration. Inversely, IER technologies are preferred over SX even at uranium concentrations in the pregnant solution above 500 ppm in cases where it is not feasible to achieve a good phase separation between solvent and pregnant solution. Also, IER is the preferred technology to recover uranium from alkaline leach liquors.

The flowsheet of uranium recovery with solvent extraction is outlined in figure 6.2. In SX, the clarified acid leach liquor is brought into contact with the organic phase, which consists of the extractant, a solvent modifier to achieve a better phase separation and to avoid the formation of a third phase, and a diluents, usually kerosene. As extractant is used frequently a weak amine such as Alamine® 336 (Cognis Corporation), a tri-octyl/decyl amine.

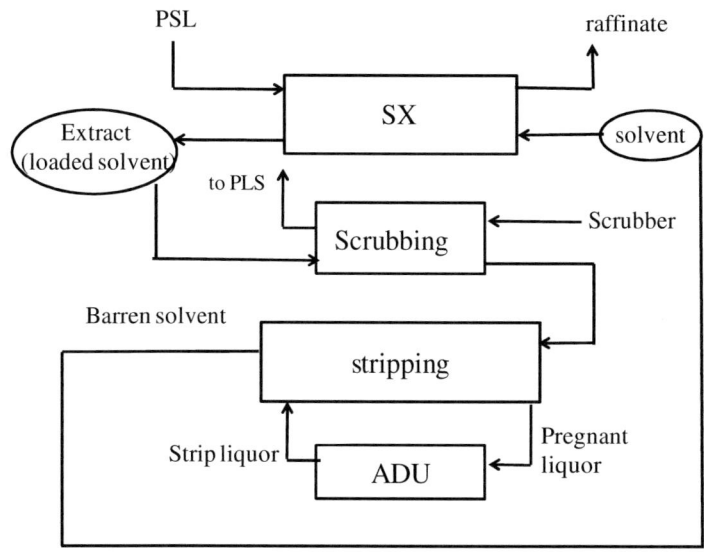

Figure 6.2 uranium recovery with SX

Uranium is loaded on the solvent as anionic uranyl sulfate complex:

$2\ (R_3NH)_2SO_{4\ o} + UO_2^{2+}{}_{aq} + 2\ HSO_4^-{}_{aq} \leftrightarrows (R_3NH)_4UO_2(SO_4)_{3o} + H_2SO_{4aq}$

$_o$ and $_{aq}$ indicate the organic and aqueous phases respectively.

Previously, the amine extractant is protonated using an acid:
$2\ R_3N_o + H_2SO_{4aq} \leftrightarrows (R_3NH)_2SO_{4o}$
$(R_3NH)_2SO_{4o} + H_2SO_{4aq} \leftrightarrows 2\ (R_3NH)HSO_{4o}$

Uranium is recovered by stripping with a suitable stripping agent, such as: Na_2CO_3, $NaCl+0.05M\ H_2SO_4$, $NaNO_3+HNO_3$ or $(NH_4)_2SO_4$ with NH_4OH to adjust the pH to slight acidic where uranium is stripped under carefully controlled pH. If the use of

NH_4OH can lead to effluent problems or if the solvent has low capacity and therefore there is high loading with HSO_4^-, strong acid such as H_2SO_4 can be used for stripping.

Na_2CO_3 stripping:

$(R_2NH_2)_4UO_2(SO_4)_{3o} + 7\ Na_2CO_{3aq} \longrightarrow 4\ RNH_o + Na_4UO_2(CO_3)_{3aq} + 3\ Na_2SO_{4aq} + 4\ NaHCO_{3aq}$

$(R_2NH_2)_2SO_{4o} + 2\ Na_2CO_{3aq} \longrightarrow 2\ R_2NH_o + Na_2SO_{4aq} + 2\ NaHCO_{3aq}$

$(R_2NH_2)HSO_{4o} + 2\ Na_2CO_{3aq} \longrightarrow R_2NH_o + Na_2SO_{4aq} + 2\ NaHCO_{3aq}$

$(NH_4)_2SO_4$ and NH_4OH stripping:

$(R_3NH)_4UO_2(SO_4)_{3o} + 4\ NH_4OH_{aq} \leftrightarrows UO_2SO_{4aq} + 2\ (NH_4)_2SO_{4aq} + 4\ R_3N_o + 4\ H_2O$

Sulfuric acid stripping:

$(R_3NH)_4UO_2(SO_4)_{3o} + 4\ H_2SO_{4aq} \leftrightarrows 4\ (R_3NH)HSO_{4o} + UO_2SO_{4aq} + 2\ H_2SO_{4aq}$

Before stripping, if some impurities have been loaded on the extractant, a number of scrubbing steps are included to eliminate the impurities. Scrubbing can be done using acidified water or $(NH_4)_2SO_4$ with NH_4OH. After stripping, the solvent (amine) is regenerated with NH_4OH to convert it to the free amine form before recycling it back to the SX circuit. There, the first step is the protonation followed by uranium extraction and stripping.

Uranium is recovered from the pregnant strip solutions by precipitation with NH_4OH or $NaOH$ as ammonium diuranate, (yellow cake), $(NH_4)_2U_2O_7$, or sodium diuranate, $Na_2U_2O_7$. In the case of H_2SO_4 stripping, most of the H_2SO_4 in the pregnant strip solution is neutralized with lime and after filtering out $CaSO_4$, the rest is neutralized with $NaOH$ or NH_4OH to precipitate the yellow cake. There exist a process to recover uranium from the

H_2SO_4 strip pregnant solution by IX using a chelating resin (Rezkallah and Dunn, 2015). The chelating resin is a phsphonic type, either AMP type or Diphonix® type. Regeneration is done with Na_2CO_3 solution and uranium is recovered by precipitation. This process is discussed later (page 170).

If the ore is rich in lime and consumes large quantities of H_2SO_4 then carbonate leaching is practiced. Carbonate leach is more selective for uranium than acid leach and the carbonate leach solution contains fewer impurities. In fact in some cases, uranium can be obtained from the leach liquors by direct precipitation as sodium diuranate using sodium hydroxide. However, one deficiency of this process is incomplete uranium precipitation. Even though the barren solution is recarbonated and recycled back to the leaching step, still some solution losses take place. Also, from ores such as carnotite where uranium is found together with vanadium, these two elements co-precipitate resulting into a contaminated yellow cake. Thus, ion exchange offers the possibility of producing purer yellow cake and with fewer uranium losses.

After leaching, the pregnant lixiviant is separated from the remaining solids and is further processed to recover uranium. If ion exchange technology is used, the suspended solids content of the pregnant solution will determine the ion exchange equipment to be used, whether this will be a fixed bed, fluidized bed or moving packed bed continuous ion exchange or Resin-In-Pulp (RIP) system.

Complex-ion equilibrium

In chapter 4, the equilibria considered involved simple ions. In this paragraph we consider the case where one of the ions can form complexes (Helfferich, 1962, p.202-222). The formation of complexes can significantly affect the selectivity of an IER. Take as an example a SBA resin in the SO_4^{2-} form in contact with a solution of UO_2SO_4. The UO_2^{2+} in solution would be excluded because of the Donnan potential. However, UO_2^{2+} forms complexes with the SO_4^{2-} ions which are not excluded and can penetrate into the resin where they can be fixed in exchange for the SO_4^{2-} ions thus greatly increasing the selectivity of the resin. The uranyl and the sulfate ions can form the following complexes (Ahrland, 1951):

$$UO_2^{2+} + SO_4^{2-} \leftrightarrows [UO_2SO_4] \qquad K_1 = 10^{2.98} \qquad (6.1)$$

$$UO_2^{2+} + 2\,SO_4^{2-} \leftrightarrows [UO_2(SO_4)_2]^{2-} \quad K_2 = 10^{2.83} \qquad (6.2)$$

$$UO_2^{2+} + 3\,SO_4^{2-} \leftrightarrows [UO_2SO_4)_3]^{4-} \quad K_3 = 10^{3.45} \qquad (6.3)$$

where K_1, K_2 and K_3 are the cumulative stability constants of the corresponding complexes.
We also have:

$$SO_4^{2-} + H^+ \leftrightarrows HSO_4^- \qquad K_4 = 1/K_d = 8.3*10 \qquad (6.4)$$

The corresponding complex-ion equilibria are given below:

$$([UO_2SO_4]) = K_1 *(UO_2)*(SO_4) \qquad (6.5)$$

$$([UO_2(SO_4)_2]) = K_2 *(UO_2)*(SO_4)^2 \qquad (6.6)$$

$$([UO_2SO_4)_3]) = K_3 *(UO_2)*(SO_4)^3 \qquad (6.7)$$

where ( ) denote molar concentrations. The total concentration of uranium in solution, (U), is given by:

$$(U) = (UO_2) + ([UO_2SO_4]) + ([UO_2(SO_4)_2]^{2-}) + ([UO_2SO_4)_3]^{4-}) \qquad (6.8)$$

substituting eqs (6.5), (6.6) and (6.7) into (6.8) we obtain:

$$(U) = (UO_2)*[\ 1 + K_1 *(SO_4) + K_2 *(SO_4)^2 + K_3 *(SO_4)^3\ ]$$

or:

$$(U) = (UO_2)\ \Sigma\ (K_n\ (SO_4)^n) \qquad \text{for n=0 to 3} \qquad (6.9)$$

the fraction x_j of a species with coordination number j is given by:

$$x_j = ([UO_2(SO_4)_j]) / (U) \qquad (6.10)$$

and substituting (6.9) into (6.10) and using eq. 6.5-6.7 we obtain:

$$x_j = K_j\ (SO_4)^j / \Sigma\ (K_n\ (SO_4)^n) \qquad \text{for n=0 to 3} \qquad (6.11)$$

equation (6.11) shows that the fractions x_j depend on the concentration of the free SO_4^{2-} ions in solution, (SO_4). At low SO_4 concentration, the complexes with low coordination number are favored. As (SO_4) increases, the higher coordination number complexes are favored. Thus, each fraction x_j passes through a maximum at a given concentration of SO_4. Figure 6.3 illustrates the speciation of uranium as a function of (SO$_4$), as derived from eq. (6.11).

The independence of the fractions xj from the UO$_2$ concentration is valid as long as the total UO$_2$ concentration is significantly lower than the total anion concentration in which case the free sulfate concentration becomes comparable to the total anion concentration.

The above complexes can now combine with the ion exchange equilibria:

$$[UO_2(SO_4)_2]^{2-} + R_2\text{-}SO_4 \leftrightarrows R_2\text{-}[UO_2(SO_4)_2] + SO_4^{2-}$$

with selectivity coefficient $K_{(UO2(SO4)2)/SO4)}$

$$[UO_2(SO_4)_3]^{4-} + 2\,R_2\text{-}SO_4 \leftrightarrows R_4\text{-}[UO_2(SO_4)_3] + 2SO_4^{2-}$$

with selectivity coefficient $K_{(UO2(SO4)3/SO4)}$

$$UO_2^{2+} + n\,SO_4^{2-} \leftrightarrows [UO_2(SO_4)_n]^{2-2n}$$

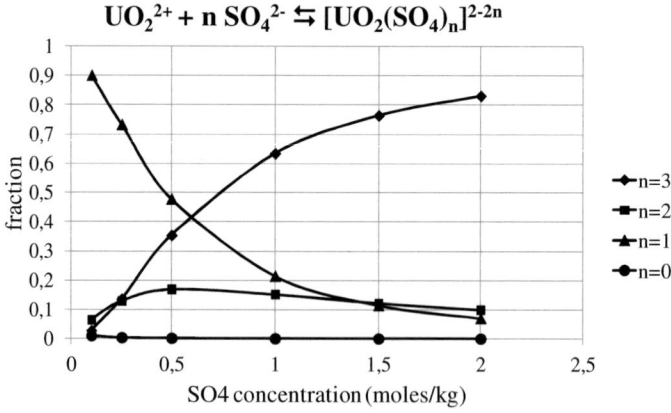

Figure 6.3 Uranyl sulfate speciation

and applying the law of mass action,

$$(R_2\text{-}[UO_2(SO_4)_2]) = K_{(UO2(SO4)2)/SO4)} * ([UO_2(SO_4)_2]) * (R_2\text{-}SO_4) / (SO_4) \quad (6.12)$$

and

$$(R_4\text{-}[UO_2(SO_4)_3]) = K_{(UO2(SO4)3/SO4)} * ([UO_2(SO_4)_3]) * (R_2\text{-}SO_4)^2/(SO_4)^2 \quad (6.13)$$

substituting eqs (6.6) and (6.7) in eqs (6.12) and (6.13) we obtain:

$$(R_2\text{-}[UO_2(SO_4)_2]) = K_{(UO2(SO4)2)/SO4)} * (K_2 *(UO_2)*(SO_4)) * (R2\text{-}SO_4)$$

and

$$(R4\text{-}[UO_2(SO_4)_3]) = K_{(UO2(SO4)3/SO4)} * (K_3 *(UO_2)*(SO_4)) * (R2\text{-}SO_4)^2$$

the total uranium concentration in the resin, (U)res, would then be:

$$(U)res = (R_2\text{-}[UO_2(SO_4)_2]) + (R_4\text{-}[UO_2(SO_4)_3])$$

$$= (UO_2)(SO_4)[\, K_{(UO2(SO4)2)/SO4)}*K_2 * (R2\text{-}SO_4) + K_{(UO2(SO4)3/SO4)}* K_3 * (R2\text{-}SO_4)^2] \quad (6.14)$$

A simplification of the above can be obtained as follows. Since the concentration of SO_4 in the resin is high, we can assume that the predominant species in the resin is the one with the highest coordination number, that is eq (6.14) becomes:

$$(U)res = (UO_2)(SO_4)\, K_{(UO2(SO4)3/SO4)} * K_3 * (R_2\text{-}SO_4)^2 \quad (6.15)$$

From eq. (6.11), as also seen in figure 6.3, at the SO_4 concentration usually found in acid leach liquors (about 25 g SO_4/L or about 0.25 moles/kg) the predominant species in the solution is the neutral complex $[UO_2SO_4]$. We then have:

$$([UO_2SO_4]) = (U)sol.$$

Substituting eq. (4.16) in (4.26) we obtain:

$$(U)res / (U)sol = \frac{K_{(UO2(SO4)3/SO4)} * K_3 * (R2\text{-}SO_4)^2}{K_1} \quad (6.16)$$

The ratio (U)res / (U)sol is also called distribution coefficient.

In the same way, the mole fractions of the uranyl complexes in solutions of UO_2^{2+} in varying concentration of H_2SO_4 can be calculated using equations 6.1 to 6.4. The independence of the fractions x_j from the UO_2 concentration is valid as long as the total UO_2 concentration is significantly lower than the total anion concentration. In this case, where $[SO_4^{2-}]_{total} > U(VI)$, at a pH of about 1.0, uranium is found mainly as the neutral or divalent anionic form (Henning *et al*, 2007).

7. Acid leaching

Chemistry of acid leach

Uranium ores contain in general both, hexavalent and tetravalent uranium. The hexavalent is soluble in H_2SO_4 while the tetravalent is insoluble and is leached in the presence of an oxidizer. As oxidizer, Fe^{3+} is used and the resulting Fe^{2+} is oxidized back to Fe^{3+} using an oxidant such as MnO_2:

$$UO_2 + 2\ Fe^{3+} = UO_2^{2+} + 2\ Fe^{2+}$$
$$2\ Fe^{2+} + MnO_2 + 4\ H^+ = 2\ Fe^{3+} + Mn^{2+} + 2\ H_2O$$

A pregnant solution in H_2SO_4 leaching contains typically: UO_2^{2+}, Fe^{2+}, Fe^{3+} SO_4^{2-}, Cl^-, SiO_2, as well as a variety of other metals, such as VO^{2+}, Fe^{2+}, Mn^{2+}, Co^{2+}, Ni^{2+}, Cu^{2+}, Zn^{2+} which are found in cationic form and which do not interfere with the uranium absorption by the resin (Preuss and Kunin, 1958). Other metals that form anionic sulfate complexes and therefore removed by the resin along with uranium are iron (III), Fe^{3+}, and thorium, Th^{4+}. If vanadium is found in the pentavalent state, then it is found in the form of vanadate, VO_4^{3-}, or metavanadate,

VO^{3-}, which can also be fixed on the anion exchange resin. Similarly, molybdates, $[MoO_4]^{2-}$, are also removed along with uranium.

A typical analysis of the pregnant solution may be as follows:

U	50-300 mg U_3O_8 /L
Fe^{3+}	0.5 – 4 g/L
SO_4^{2-}	20-25 g/L
Cl^-	0.1- 1 g/L
SiO_2	0.1 – 1 g/L
pH	1.5 to 2.2

Uranium in this solution is found as the cation UO_2^{2+}. In presence of SO_4^{2-} and at a pH<2 it forms the complexes:

$$UO_2^{2+} + SO_4^{2-} \leftrightarrows [UO_2SO_4] \quad (7.1)$$
$$UO_2^{2+} + 2\,SO_4^{2-} \leftrightarrows [UO_2(SO_4)_2]^{2-} \quad (7.2)$$
$$UO_2^{2+} + 3\,SO_4^{2-} \leftrightarrows [UO_2(SO_4)_3]^{4-} \quad (7.3)$$

We also have the equilibrium SO_4^{2-} - HSO_4^-:

$$SO_4^{2-} + H^+ \leftrightarrows HSO_4^- \quad (7.4)$$

Iron (III) forms complexes with SO_4^{2-}:

$$Fe^{3+} + SO_4^{2-} \leftrightarrows [FeSO4]^+$$
$$Fe^{3+} + 2\,SO_4^{2-} \leftrightarrows [Fe(SO4)_2]^- \quad (7.5)$$

In view of these complexes formed in acid leach liquors, uranium is removed from the solution using anion exchange resins, either strong or weak base, most frequently strong base. At saturation, ionic sites of the resin will be occupied by HSO_4^-, SO_4^{2-}, $Fe(SO_4)_2^-$, $[UO_2(SO_4)_2]^{2-}$ and $[UO_2(SO_4)_3]^{4-}$. Iron (II) does not form sulfate complexes and therefore is not fixed by the anion exchange resin.

The ion exchange reactions of loading uranium on the SBA resin can be presented by :

$$2\ R\text{-}X + SO_4^{2-} \leftrightarrows R_2\text{-}SO_4 + 2\ X^- \tag{7.6}$$

$$R\text{-}X + HSO_4^- \leftrightarrows R\text{-}HSO_4 + X^- \tag{7.7}$$

$$[UO_2(SO_4)] + 2\ R_2\text{-}SO_4 \leftrightarrows R_4\text{-}[UO_2(SO_4)_3] \tag{7.8}$$

$$2\ Fe(SO4)_2^- + R_2\text{-}SO_4 \leftrightarrows 2\ R\text{-}[\ Fe(SO_4)_2^-] + SO_4^{2-} \tag{7.9}$$

where X is the initial ionic form of the resin after elution.

It should be noted that uranium can be fixed from acid leach liquors on a SAC resin as the cation UO_2^{2+} (Ford, 2009). However, SAC resins are not particularly selective for UO_2^{2+} and the resin fixes other cations as well.

Absorption of uranium

The average composition of a pregnant solution in in situ leaching (ISL) is as follows, in g/L (IAEA, 2001):

U: 0.015-0.1
H_2SO_4: 1-7
pH=1.2-2
Fe^{2+}: 0.2-1.5
Fe^{3+}: 0.15-0.9
Ca^{2+}: 0.3-0.6
Mg^{2+}: 0.3-1.6
SO_4^{2-}: 10-25
Cl^-: 0.2-1.7
NO_3^-: 0.06-0.6
Al^{3+}: 0.3-2.5
SiO_2: 0.05-0.5

The concentration of SO_4^{2-} in a pregnant solution is, as indicated above, 10-25 g SO_4^{2-} /L, or 0.10-0.25 moles/kg. As figure 4.7 suggests (page 148), the prevailing species of uranium in this solution is the neutral complex [UO_2SO_4] (Arden and Wood, 1956 ; Chanda and Rempel, 1992). When an anion exchange resin in SO_4 form comes in contact with a pregnant solution, the neutral complex [UO_2SO_4] is not excluded by the Donnan potential and is to a large extent sorbed by the resin. Once inside the resin and due to the high SO_4^{2-} concentration since the resin is in the SO_4^{2-} form, the neutral complex [UO_2SO_4] forms the trisulfate complex on the resin sites (eq. (7.8) above). This partly explains why the anion exchange resins have a high selectivity for uranium in presence of high SO_4^{2-} concentration in the solution.

At high free acid content, the high HSO_4^- concentration will compete for the ionic sites of the resin which then will be occupied more by HSO_4^- and this will have the result that the formation of the tetravalent complex *inside the resin*, eq. (7.8), will be hindered with the loading capacity of the resin significantly reduced.

Styrenic SBA resins have the highest selectivity and are the resins mostly used in uranium recovery. Pyridine based resins have been introduced early (DAlelio, 1952) and recommended for uranium recovery (Greer et al, 1958). Since then more commercial pyridine resins have been developed for uranium recovery. This type of resins were also used in the former Soviet Union countries (anionit AMΠ)[1]. Poly(4-vinylpyridine) weak base anion exchangers and quaternized to form strong base anion exchangers have some differences with respect to styrenic resins as discussed below (page 162).

Acrylic resins have lower selectivity and formophenolic resins even lower and these resins have not found yet use in uranium recovery.

All types of anion exchange resins, weak or strong, can be used to recover uranium from acid leach solutions. The use of WBA resins was motivated by the expected higher selectivity for uranyl sulfate over ferric sulfate complex compared to SBA resins and by a more economical elution due to the fact that the remaining eluent anions on the resin in the case of Cl^- or NO_3^- elution could be quantitatively eluted with NH_4OH wash thus allowing eluent recovery. In practice however, the recovery of the eluent with NH_4OH wash did not work properly due to contamination of the resin with SiO_2 which was also eluted with the NH_4OH (Haines *et al*, 1975; Haines, 1977). Also, the kinetics of loading were found to be slower than with gel-type SBA resins. Some uranium mines in North America used WBA resins in the years 60's but today SBA resins are used in practically all

---

[1] Not to confuse with aminomethylphosphonic (AMP) resins

uranium mines. However, it remains to be seen whether with pregnant solutions containing high iron content, WBA resins present an overall advantage over SBA resins.

MR type SBA resins have also been introduced in the last years. MR resins show different kinetics than gel types due to the larger surface area that the MR resins have. This difference in kinetics is seen especially in the elution of uranium, possibly amplified due to the large size of the complexes involved.

The saturation capacity of SBA resins depends upon a number of parameters, both due to resin properties (particle size, moisture) and to solution characteristics (uranium concentration, chlorides or nitrates content, iron, pH) as it is discussed below.

Effect of resin particle size

As discussed in Chapter 5, when the kinetics of ion exchange is particle diffusion controlled and the diffusion coefficient is uniform throughout the resin bead, the effect of resin particle size on the breakthrough curves is represented in figure 5.16 (page 123). It has been observed however that with certain gel-type SBA resins having high degree of crosslinking (low moisture, in the range of 42-45%) the breakthrough curves of resins having different particle size but similar total exchange capacities and moisture content do not show the pattern of fig. 5.16 but they are shifted to each other as shown in figure 7.1.

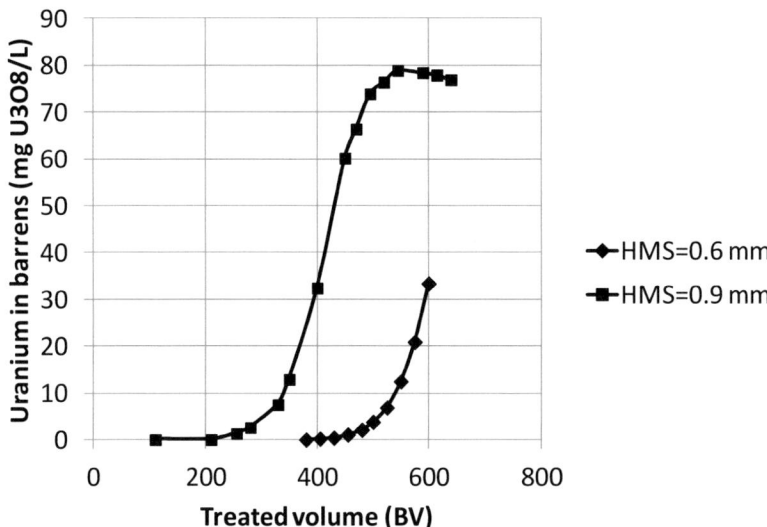

Figure 7.1 Effect of particle size on breakthrough curves for a gel-type SBA

In figure 7.2 it is shown the effect of particle size for a MR type SBA resin. The particle size effect in the case of MR resins having a larger internal surface area, indicates that the diffusion of the sulfate complexes is uniform throughout the resin beads during the loading period. For the gel-type resin (figure 7.1) it appears that the kinetic behavior is similar to a core-shell type where the particle diffusion slows down as the loading of the resin increases.

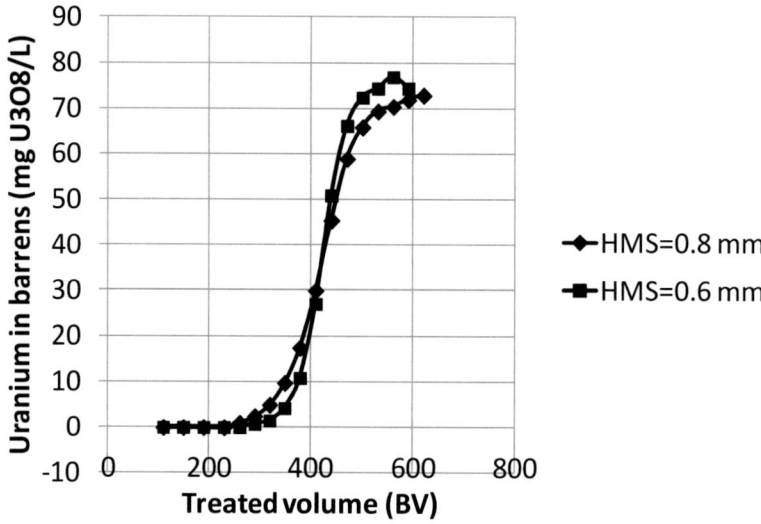

Figure 7.2 Effect of particle size on breakthrough curves for an MR type SBA resin.

This results into an early breakthrough and, as figure 7.1 shows, a decreased "saturation" capacity of the resin before the inner part of the resin becomes exhausted. This behavior has not been observed with all gel-type resins. Gel-type resins having lower degree of crosslinking (higher moisture) show this behavior to a lesser extent and they approach the behavior of MR resins. The high moisture resins however have lower total exchange capacity compared to low moisture resins. It remains therefore to be determined on a case by case basis which type of resin would give a higher capacity and a lower uranium leakage under given conditions. In addition to that, one should consider also the different rate of silica fouling of the two types of resin (Chapter 10) as well as a better elution of the high moisture resins (see below).

Effect of uranium concentration

At similar ionic background of the solution, the higher the uranium concentration in solution, the higher the saturation capacity of the resin. This is illustrated in the isotherm of figure 7.3. In this curve, a solution containing initially 200 mg U_3O_8/L, 22 g SO_4/L and 2.5 g Fe^{3+}/L at a pH of 1.7 was put together with various quantities of resin in flasks under agitation until equilibrium was reached (usually overnight). After equilibrium, the amount of uranium fixed by the resin and the uranium concentration in solution were determined and plotted in figure 7.3.

The values of uranium loaded on the resin in equilibrium with low uranium concentrations in solution are those that will determine the barrens concentration. Thus, when the eluted resin still contains a certain quantity of residual uranium, then the equilibrium solution concentration that corresponds to this residual uranium value will be the minimum barrens concentration that can be achieved. At the other end of the curve, the resin loading corresponding to the feed solution concentration indicates the maximum uranium load on the resin which goes to elution.

Generally speaking, the uranium equilibrium capacity is correlated to the total exchange capacity of resins of similar physical and chemical structure (Mikhaylenko and van Deventer, 2009). Resins with different structure however, for example type 2 vs type 1 SBA resins, or vinyl pyridine vs sryrenic resins, show different uranium capacities (McGarvey and Ungar, 1981).

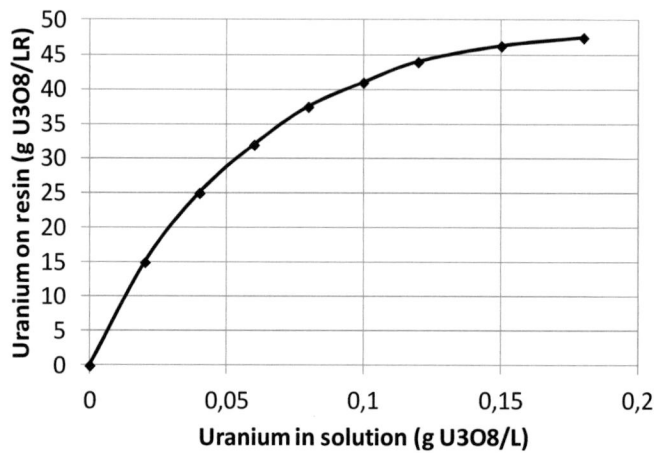

Figure 7.3 Equilibrium loading, gel type 1 SBA resin, equilibrated with a solution containing initially 22 g SO_4/L, 2.5 g Fe^{3+} at pH=1.7

Effect of pH

The pH of the leach liquor affects the operating capacity of the resin significantly. On one hand, low pH, as eq. (7.1) to (7.4) indicate, increases the ratio HSO_4^- to SO_4^{2-} and the resin sites will be occupied to a larger extent by HSO_4^- ions. At the same time, the speciation of uranium will be affected accordingly, as figure 4.3 indicates which means that when the [UO_2SO_4] complex enters the resin, the formation of the di- or tetravalent complex will be suppressed. Consequently, at saturation, less uranyl sulfate complexes will be found on the resin (see eq. 7.8). High pH favors the adsorption of iron as it will be discussed below. It

Effect of uranium concentration

At similar ionic background of the solution, the higher the uranium concentration in solution, the higher the saturation capacity of the resin. This is illustrated in the isotherm of figure 7.3. In this curve, a solution containing initially 200 mg U_3O_8/L, 22 g SO_4/L and 2.5 g Fe^{3+}/L at a pH of 1.7 was put together with various quantities of resin in flasks under agitation until equilibrium was reached (usually overnight). After equilibrium, the amount of uranium fixed by the resin and the uranium concentration in solution were determined and plotted in figure 7.3.

The values of uranium loaded on the resin in equilibrium with low uranium concentrations in solution are those that will determine the barrens concentration. Thus, when the eluted resin still contains a certain quantity of residual uranium, then the equilibrium solution concentration that corresponds to this residual uranium value will be the minimum barrens concentration that can be achieved. At the other end of the curve, the resin loading corresponding to the feed solution concentration indicates the maximum uranium load on the resin which goes to elution.

Generally speaking, the uranium equilibrium capacity is correlated to the total exchange capacity of resins of similar physical and chemical structure (Mikhaylenko and van Deventer, 2009). Resins with different structure however, for example type 2 vs type 1 SBA resins, or vinyl pyridine vs sryrenic resins, show different uranium capacities (McGarvey and Ungar, 1981).

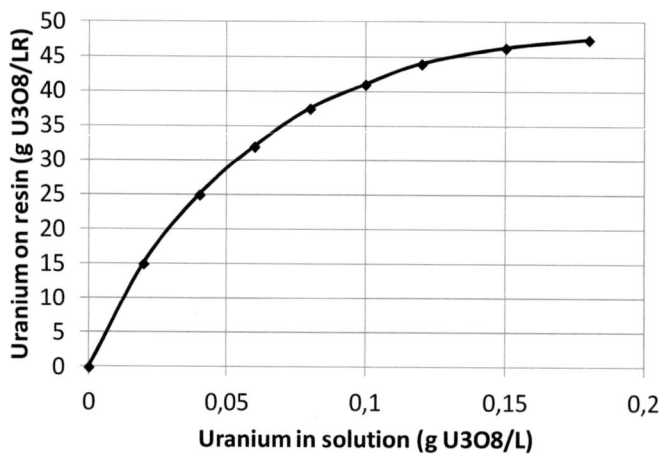

Figure 7.3 Equilibrium loading, gel type 1 SBA resin, equilibrated with a solution containing initially 22 g SO_4/L, 2.5 g Fe^{3+} at pH=1.7

Effect of pH

The pH of the leach liquor affects the operating capacity of the resin significantly. On one hand, low pH, as eq. (7.1) to (7.4) indicate, increases the ratio HSO_4^- to SO_4^{2-} and the resin sites will be occupied to a larger extent by HSO_4^- ions. At the same time, the speciation of uranium will be affected accordingly, as figure 4.3 indicates which means that when the $[UO_2SO_4]$ complex enters the resin, the formation of the di- or tetravalent complex will be suppressed. Consequently, at saturation, less uranyl sulfate complexes will be found on the resin (see eq. 7.8). High pH favors the adsorption of iron as it will be discussed below. It favors the formation of $U_2O_5(SO_4)_2^{2-}$ and $U_2O_5(SO_4)_3^{4-}$ which results in higher resin capacity but at slower kinetics. However, it also affects the hydrolysis of various ionic metal impurities in solution. Therefore, there is not a simple relationship between pH and uranium saturation capacity. As an illustration only, the saturation capacity at pH=2.2 is about 15% higher than at a pH=1.8 with a feed solution containing 75 mg U_3O_8/L, 17 g/L of SO4 and 0.26 g/L of Fe^{3+}. In general, for all these reasons, the optimum pH found in practice is 1.5-2.2.

Poly(4-vinylpyridine) weak and strong base resins show somewhat better capacity than styrenic resins at a pH of about 1.5 (Chanda and Rempel, 1992) but at higher pH the capacities become similar.

Effect of chlorides and nitrates

SBA resins have high selectivities for Cl^- and NO_3^- ions and therefore, the presence of these ions in the pregnant solution results into a decreased capacity of the resin for uranium. However, the decrease in breakthrough and saturation capacity due to the presence of chlorides is much higher than one would expect from only the selectivity coefficients of the resin. For example, a gel-type SBA resin, with a solution containing 75 mg U_3O_8/L, 0.24 eq/L Cl-, 0.11 eq/L SO_4 (a total of 0.35 eq/L of Cl-+ SO_4) and a pH of 2.2 shows a loading capacity of only 20% of that obtained with a solution containing 75 mg U_3O_8/L, 0.35 eq/L of SO_4 and a pH of 2.2. Still, the selectivity of the resin for SO_4^{2-} is higher than that for Cl^-.

The big effect of chlorides, at the 8.5 g/L level, on the resin capacity for uranium is possibly due to the fact that when chlorides occupy significant part of the ionic sites of the resin, the formation of the R_4-$[UO_2(SO_4)_3]$ complex on the resin, where four close distanced functional groups fix the quadrivalent $[UO_2(SO_4)_3]^{4-}$, does not take place as easily with the consequence that the affinity of the resin for other uranylsulfate complxes such as $[UO_2(SO_4)_2]^{2-}$ drops. In other words, the presence of Cl^- interferes on the selectivity of the resin for $[UO_2(SO_4)_3]^{4-}$ over SO_4^{2-} ions. This selectivity varies according to the degree of loading of the SO_4 ions on the resin. The less SO_4 ions are fixed on the resin (and therefore more Cl^-), the lower is the selectivity for $[UO_2(SO_4)_3]^{4-}$. Poly(4-vinylpyridine) weak and strong base resins are less affected by increasing chlorides concentration compared to styrenic resins (Chanda and Rempel, 1992).

For PLS with high Cl- content, solvent extraction using tertiary amines faces also a big capacity loss.

Recently, selective resins of IDA or AMP type resins have been employed to recover uranium from acid leach liquors containing high Cl^- concentrations (Rezkallah, 2012; Carr *et al*, 2012; Ogden and Soldenhoff, 2013), thus avoiding the use of SX in these cases. According to this approach, pregnant leach solutions containing up to 20 g/L of chlorides can successfully be treated with AMP type resins giving operating capacities of 20 g U/L_R at flow rates of 2.5 BV/h and 15 g U/LR at 5 BV/h. Elution was done first with NH_4OH to neutralize the resin sites in the H^+ form followed by Na_2CO_3 to elute the uranium. Iron however is a problem and needs to be removed separately.

A similar approach was taken in solvent extraction using phosphorous-containg functionalities liquids. Thus, solvents containing D2EHPA and Alamine® 336 (Soldenhoff *et al*, 2000) or D2EHPA and Cyphos IL-101 (trihexyl(tetradecyl)phosphonium chloride) (Zhu *et al,* 2014) or a mixture of D2EHPA, tertiary amine and tri-butyl phosphate (Ballestrin et al, 2014).

Effect of iron

Iron constitutes the major impurity in acid leach solutions. Although Fe^{2+} does not form a sulfate complex, Fe^{3+} forms complexes such as $Fe(SO_4)_2^-$ which are fixed on the resin together with uranium. The selectivity of the resin for ferric iron is lower than for the uranyl complexes but pH has an important effect. Above pH 1.8, the selectivity of the resin for ferric iron increases significantly. Even though the resin prefers uranium over iron (III), increasing concentration of Fe^{3+} in the leach solution, at the same uranium concentration, has the consequence that at saturation, the resin will contain more Fe^{3+} and less uranium and therefore the loading capacity of the resin for uranium will be reduced. For example, at a level of 4 g Fe^{3+}/L and a pH of 1.8 the saturation capacity of a gel type SBA resin is decreased by about 15% compared to that at Fe^{3+} levels below 0.5 g/L.

Different anion exchange resins have different selectivities for U over Fe^{3+}. Strong base anion exchange resins are more selective for $Fe(SO_4)_2^-$ than weak base resins and in that respect, WBA resins are similar to the tertiary amines liquid extractants in SX. Poly(4-vinylpyridine) weak and strong base resins show somewhat higher affinity for uranium over $Fe(SO_4)_2^-$ (Chanda and

Rempel, 1992). Conventional SBA resins also vary in U/Fe^{3+} selectivities. Thus, gel type SBA resins are less selective for U/Fe^{3+} than MR resins. For example, with a PLS with 200 ppm U and 2.5 g/L Fe^{3+}, the ratio of U/Fe^{3+} (expressed in g/LR) at saturation is about 10 for gel-type and 30 for MR type SBA resins.

Thorium

Thorium forms anionic sulfate complexes:

$$Th^{4+} + 3\ SO_4^{2-} \leftrightarrows [Th(SO_4)_3]^{2-},$$

which can be fixed on anion exchange resins along with uranium. The thorium complex is fixed less strongly than uranyl sulfate and is displaced by uranium, in a similar way as ferric sulfate complexes. Thorium can be recovered separately from uranium on a three-column system containing the same anion exchanger where uranium is recovered from the first column and uranium-free thorium from the third column (Arden et al, 1959).

Molybdenum

Molybdenum occurs in most uranium minerals and can end-up in the leach liquors in both, acid and alkaline leaching. Molybdenum in the VI state can form complexes, as uranium does, of the $MoO_2(SO_4)_n^{2-2n}$ type,. These complexes however, are more strongly fixed by SBA resins than uranyl complexes and consequently, they tend to accumulate on the resin, thus causing a reduction in operating capacity of the resin for uranium.

Once on the resin, molybdenum can be removed by 5-10% NaOH (see page 221, resin fouling).

Vanadium

Depending on the amount of oxidant used, vanadium can be present as quadravalent, VO^{2+} or pentavalent, VO_3^- or VO_4^{3-}. In the +5 state therefore, vanadium is fixed on the SBA resin along with uranium. In order not to contaminate uranium with vanadium, the redox potential could be adjusted so that vanadium is found in the +4 state as cation.

Elution

The uranium complex can be eluted from the resin with an acidified nitrate or chloride solution or with H_2SO_4. The elution reactions can be presented as:

$$R_4\text{-}[UO_2(SO_4)_3] + 4\,X^- \leftrightarrows UO_2^{2+} + 3\,SO_4^{2-} + 4\,R\text{-}X \quad (7.10)$$

$$R\text{-}[Fe(SO_4)_2^-] + X^- \leftrightarrows R\text{-}X + Fe(SO_4)_2^- \quad (7.11)$$

$$R_2\text{-}SO_4 + 2\,X^- \leftrightarrows 2\,R\text{-}X + SO_4^{2-} \quad (7.12)$$

Where X^- is NO_3^-, Cl^- or HSO_4^-. Typical eluent compositions are 80 g/L NH_4NO_3 (1 eq/L) + 20 g/L H_2SO_4, 60 g/L NaCl (1 eq/L) + 20 g/L H_2SO_4 and 10-12% H_2SO_4.

At the end of the elution, the resin will be found mainly in the NO_3^-, Cl^- or HSO_4^-/SO_4^{2-} form respectively.

High temperatures (40-60°C) have a positive effect on the efficiency of elution.

From H_2SO_4 elution equilibrium curves, that is, isotherms of uranium in a solution background of 8-10% H_2SO_4 (van Deventer, 2013), a solution concentration of about 6 g/L in uranium corresponds to a resin loading of about 15-25 g/L_R uranium. This implies that the uranium concentration in the H_2SO_4 concentrated eluates would be limited to about 5-8 g/L of uranium. The concentrated eluates of a $NH_4NO_3 + H_2SO_4$ elution can contain 25-35 g U/L or even more (Patrin, 2005).

Because of the lower selectivity of the resin for iron compared to uranium, the first part of the eluate contains proportionally more iron than in the subsequent parts.

Figure 7.4 shows the elution profiles at ambient temperature, using the above three eluents at 1 BV/h for a gel-type SBA resin of a moisture content in the range 42-45%. In the same curve it is also included the elution of uranium using acidified $(NH_4)_2SO_4$ as eluent (80 g $(NH_4)_2SO_4$ /L + 20 g H_2SO_4/L). The elution profiles are sensitive to the eluent flow rate. Increasing flow rate results in broadening of the peaks and increasing of the eluent volume to bring the residual uranium on the resin down to a given level. Typical eluent flow rate is 0.5 BV/h.

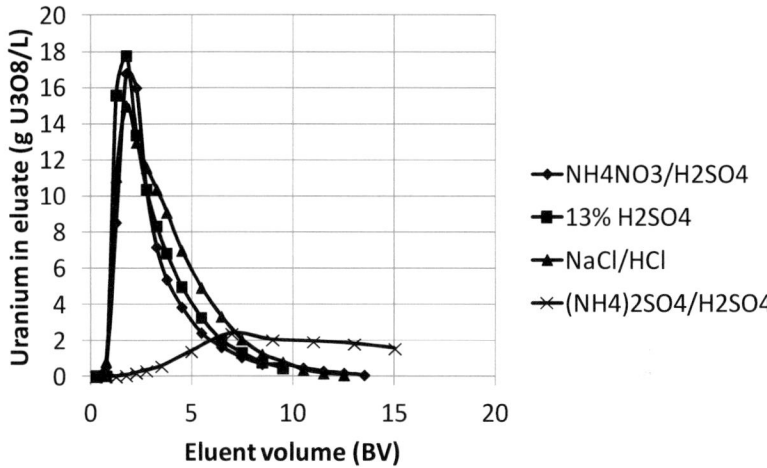

Figure 7.4 Elution profiles, SBA resin of gel type, ambient temperature.

It is interesting to note that when acidified $(NH_4)_2SO_4$ is used, at 80 g/L concentration, as eluent, the elution is very inefficient even though the selectivity of the resin for the SO_4^{2-} ions is high. In this case the elution reaction is :

$$R_4\text{-}[UO_2(SO_4)_3] + 2SO_4^{2-} \leftrightarrows 2\,R_2\text{-}SO_4 + [UO_2(SO_4)_3]^{4-}$$

As suggested in the previous section (effect of chlorides on operating capacity), the selectivity of the resin for $[UO_2(SO_4)_3]$ with respect to SO_4^{2-} is very high when the resin contains SO_4^{2-} ions to a high degree. This therefore should result into an unfavorable elution of $[UO_2(SO_4)_3]$ by SO_4^{2-} ions. This is not the case when nitrate or chloride or H_2SO_4 (HSO_4^-) elution is practiced, because as it was discussed above, the selectivity of the resin for uranyl sulfate depends on the degree of SO_4^{2-} loaded on the resin.

An analogous situation is observed with the elution of $[UO_2(CO_3)_3]^{4-}$ using HCO_3^- or CO_3^{2-} as eluent, as it will be discussed in the following section.

The effect of the specific flow rate on the elution for the NH_4NO_3 elution is illustrated in figure 7.5. The resin was an MR type SBA.

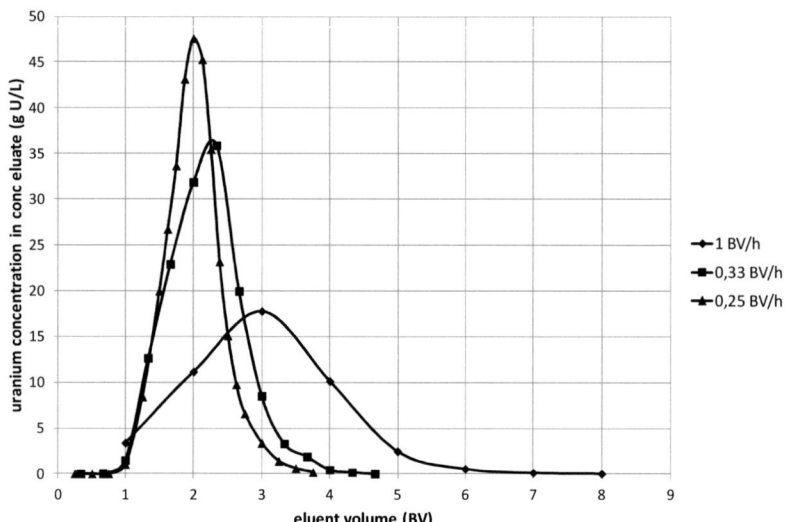

Figure 7.5 NH_4NO_3 elution, effect of the specific flow rate. MR type SBA.

By reducing the specific flow rate of the elution of a gel-type SBA resin with H_2SO_4 eluent, the effect is less pronounced. It reduces the number of BV's to reach a given residual uranium on the resin and at the same time it increases to a certain extent the concentration of the peak of the concentrated eluate. Roughly speaking, a reduction of the specific flow rate from 1 BV/h to

0.5 BV/h for a gel-type, low moisture resin, results in about 30% decrease in the number of BV's of eluate to elute better than 98% of the uranium.

For a macroporous SBA resin of RIP particle size grade, the effect of the specific flow rate on the elution with H_2SO_4 was found to be small (Udayar et al, 2011).

The main impurity of the loaded resin is iron. Even though uranium is preferentially fixed over iron, the loaded resin, in equilibrium with the feed solution, will always contain some iron. The ratio uranium to iron on the resin will depend on the selectivity coefficients of the resin, some resins having in fact low selectivity for iron. In order to increase the uranium to iron ratio on the resin before the elution step, a scrubbing step can be included using a dilute H_2SO_4 solution or part of the concentrated eluate, as the scrub solution. In this way, impurity ions are displaced and the ratio uranium to iron will be increased. Usually, when eluate is used as the scrub solution, the ratio uranium to iron increases more than with dilute H_2SO_4 solution and in addition, the uranium loading on the resin increases resulting into higher concentration of uranium in the eluate.

The eluate of the NO_3^- or Cl^- elution contains, as reaction 7.10 indicates, except UO_2^{2+} and the unused NO_3^- or Cl^-, SO_4^{2-} ions originating from the uranyl sulfate complex and the HSO_4^- and SO_4^{2-} fixed by the resin or present in the fresh eluent. If therefore after neutralization and precipitation of uranium, the barren eluate is recycled to recover the eluent, the recycled eluate will contain a certain concentration of SO_4^{2-} which will build up after each recycling. In that case, the removal of the SO_4^{2-} from the eluate, for example with lime, should be done periodically. This is discussed further in page 192.

Low moisture gel-type resin implies high total volume capacity and high crosslinking density. High moisture implies low total volume capacity and low crosslinking density. The elution profile of the high moisture gel resin is sharper than that of the low moisture resin. This is probably due to the faster kinetics of the high moisture resin (lower crosslinking density) but also due to the lower loading capacity of this resin (due to the low total volume capacity). The elution profile of the MR resin is sharper than low moisture gel type resins, possibly again, as with high moisture gel resins, due to the faster kinetics (porous structure) and lower total exchange capacity. Figure 7.6 illustrates the elution curves of an MR-type resin and a standard gel, type 1 SBA resin even though the gel-type resin in figure 7.6 has a smaller particle size than the MR type. Sharper elution profiles of MR resins compared to gel were also reported in the literature (McGarvey and Ungar, 1981; McGarvey and Hauser, 1982).

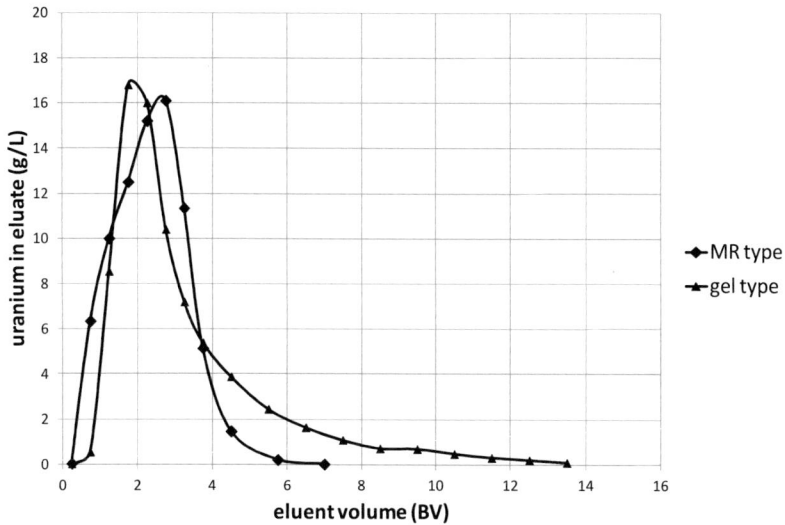

Figure 7.6 Elution of gel and MR type SBA resins with H_2SO_4 (Courtesy of The Dow Chemical Company)

Recovery of uranium from H_2SO_4 eluates

When sulfuric acid is used as eluent, uranium is removed from the eluate with solvent extraction (Eluex or Bufflex process). Alternative technologies to recover uranium from H_2SO_4 eluates include nanofiltration (NF) (Naidoo *et al*, 2015), phosphonic chelating resins (Rezkallah and Dunn, 2015) and strong acid cation exchange resins (Bester *et al*, 2016). A NF based acid recovery installation has been installed at the Kayelekera Uranium Mine in Malawi (Peacock *et al*, 2016).

The use of phosphonic resins to recover uranium (Rezkallah and Dunn, 2015) is illustrated in figure 7.7:

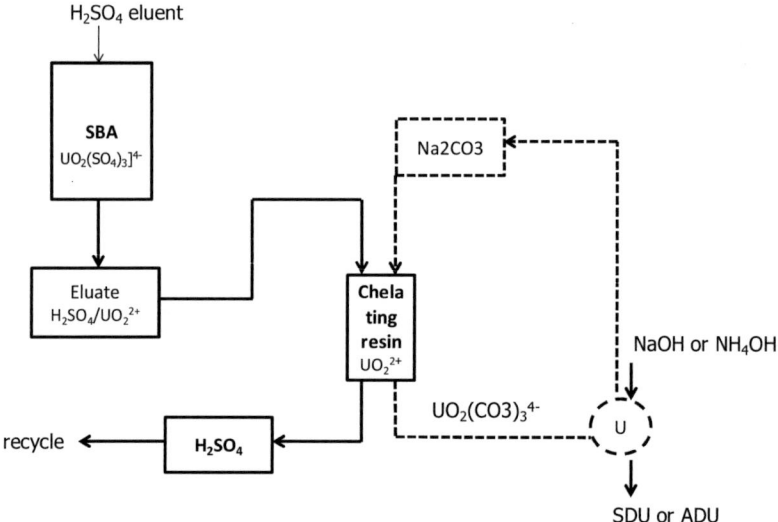

Figure 7.7 Uranium recovery from H_2SO_4 eluates using phosphonic chelating resins. Elution with Na_2CO_3.

The resin forms strong complexes with uranium and it recovers it from the H_2SO_4 solution. Elution is done with Na_2CO_3 with which uranium forms even stronger complex. In order to avoid CO_2 gas to be formed during the elution with Na_2CO_3, according to the reaction:

$$2\ R\text{-}SO_3\text{-}H^+ + Na_2CO_3 \rightarrow 2\ R\text{-}SO_3\text{-}Na^+ + CO_2\uparrow + H_2O$$

an alkaline wash can be done before the Na_2CO_3 elution, in order to neutralize the sites of the resin in the H^+ form. This however, has the result that the yield of uranium elution is slightly decreased (Dunn and Teo, 2015).

After precipitation of uranium, the Na_2CO_3 is recycled. Since the resin works in the H^+ form, some H_2SO_4 is needed to convert the resin back to the H^+ form.

The other IX approach (Bester *et al*, 2016) uses a SAC resin to recover uranium from spent H_2SO_4 eluates. In solutions such as the H_2SO_4 eluates, with high ratio $[SO_4]_{total}$ to UO_2^{2+}, uranium is found mainly as the neutral complex with minor constituent UO_2^{2+}. From H_2SO_4 eluates, and due to the displacement of the equilibrium, UO_2^{2+} is fixed on a SAC resin according to the equation:

$$2\ R\text{-}SO_3^-H^+ + UO_2^{2+} \leftrightarrows (R\text{-}SO_3^-)_2UO_2^{2+} + 2\ H^+ \qquad (7.14)$$

Since here we have a di-monovalent equilibrium, the equilibrium isotherms depend on the total solution concentration and the more dilute is the solution, the more the resin will fix the divalent ion, here the UO_2^{2+}. Therefore a dilution of the H_2SO_4 eluate favors the recovery of uranium by a SAC resin.

Once uranium has been removed from the eluate with the SAC resin, it can be eluted either with an eluent such as Na_2SO_4 or Na_2CO_3 or with HCl.

In HCl, uranium (VI) forms complexes with Cl^- according to the following reactions (Soderholm *et al*, 2011):

$UO_2^{2+} + Cl^- \leftrightarrows [UO_2Cl]^+ \qquad K_1 = 10^{1.5}$
$UO_2^{2+} + 2\ Cl^- \leftrightarrows [UO_2Cl_2] \qquad K_2 = 10^{0.8}$
$UO_2^{2+} + 3\ Cl^- \leftrightarrows [UO_2Cl_3]^- \qquad K_3 = 10^{0.4}$

Using these stability constants a speciation diagram can be constructed as illustrated in figure 7.8.

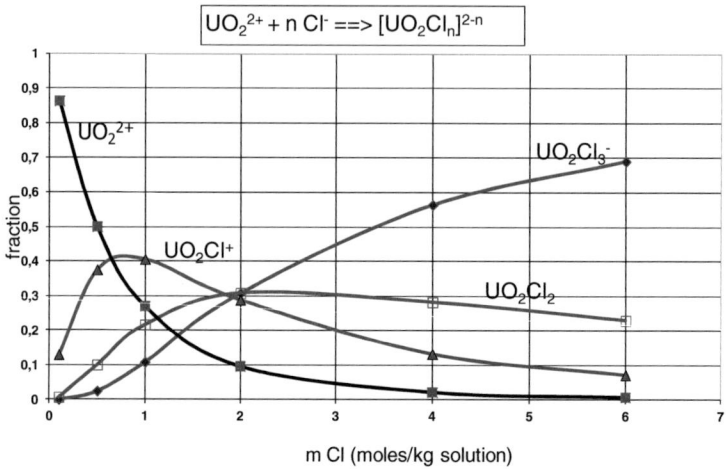

Figure 7.8. Speciation diagram for uranyl chloride complexes (stability constants taken from Soderholm *et al*, 2011).

In the elution with HCl, at a Cl⁻ concentration of ≈ 4 N, there is significant portion of uranium in an anionic complex form. After the SAC resin, the spent HCl eluate is allowed to pass through a SBA resin in the Cl⁻ form where uranium is retained while the excess HCl is recycled. Subsequently, uranium is eluted from the SBA resin with water.

The corresponding reactions are the following:

$(R-SO_3^-)_2 UO_2^{2+} + 3\ HCl \leftrightarrows UO_2Cl_3^- + 2\ R-SO_3^-H^+ + H^+$ (7.15)
$RN^+Cl^- + UO_2Cl_3^- \leftrightarrows RN^+ UO_2Cl_3^- + Cl^-$ (7.16)
$RN^+ UO_2Cl_3^- + H_2O \leftrightarrows RN^+Cl^- + UO_2^{2+} + 2\ Cl^-$ (7.17)

Overall, there is no chemical used as an eluent, except water. Of course, there is a HCl consumption during the water elution. The flowsheet of this approach is illustrated in figure 7.9.

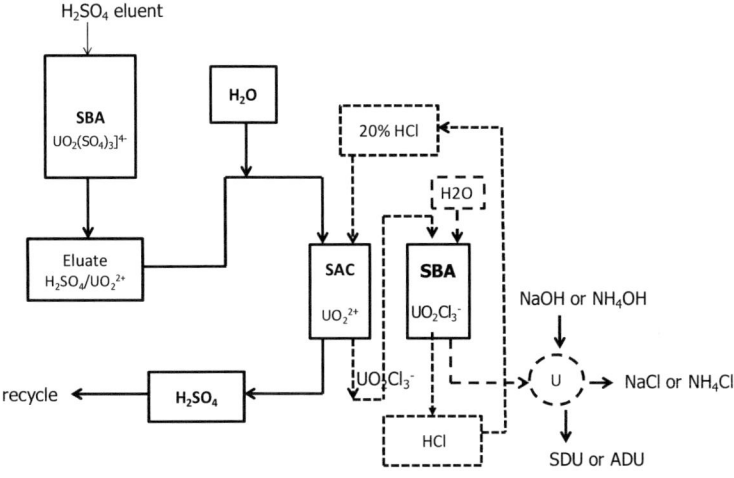

Figure 7.9 Uranium recovery from H_2SO_4 eluates using SAC resins. Elution with HCl.

Removal of uranium from mining wastes

Acid Mine Drainage

Acid mine drainage (AMD) is acidic water coming off from abandoned mines or tailings piles. Sulfide minerals, after coming in contact with air and water generate acidity from the oxidation of the metal sulfides. Depending on the mineral, metal ions dissolved in the acidic waters may be iron, copper, zinc, manganese or nickel. In AMD close to uranium mines can contain uranium to a concentration of around 10 mg/L. Anionic species are mainly SO_4^{2-} but also Cl^-, or F^- and the pH is about 2.5-3.5.

There are various treatments for AMD, the most frequent being lime precipitation. According to this method, a slurry of lime is dispersed in a tank with AMD where pH increases to about 9. At this pH the metals are precipitated as hydroxides. The slurry is then directed to a clarifier where clean water will overflow for release while settled materials (sluge) come out at the bottom. If uranium is found in the AMD, then this treatment generates radioactive sludge which needs to be disposed. Removal of uranium therefore before precipitation presents significant interest.

Because of the high sulfate concentration, uranium is found as uranyl sulfate complexes and consequently, anion exchange resins are suitable to remove and recover uranium (Ladeira and Gonçalves, 2007; Ladeira and Sicupira, 2013). Strong acid cation exchangers are not suitable due to the presence of other cations such as Ca^{2+}, Mg^{2+}, Na^+ and various metals.

In their study, (Ladeira and Sicupira, 2013), they compared strong base anion exchangers, two of them with macroporous Type 2 and one gel, Type 1 structure with two feed solutions varying in uranium, sulfates and iron contents. The criteria were loading capacity and elution characteristics. Although the results are difficult to analyze, it looks that macroporous, Type 2 resins may present some advantages over gel, Type 1 resins. Operating capacities varied depending on the resin and the feed solution composition, and were between approximately 10 and 20 g U/L_R. Split elution was done, first with 0.1 M H_2SO_4 in order to remove iron, followed by 1.5 M $NaCl+0.05$ M H_2SO_4 to elute uranium.

Contaminated water from mining sites

Removal of uranium from solutions where uranium is found in low concentrations, low ppm or ppb levels, in high ionic background and in presence of other metals, was achieved by using WBA resins impregnated with polyphenolic compounds (Rezkallah *et al*, 2016; Dow 2016). These polyphenol-impregnated WBA exchange resins were able to remove uranium from waters of a wide range of pH and ionic strength. This Dow polyphenol-impregnated WBA exchange resin was used in Diffusive Gradients in Thin Films (DGT) technique to measure uranium in waters near uranium mining sites (Drozdzak *et al*, 2016). The resin exhibited good performance in a pH range of 3-9 and an ionic background of 0.001-0.7 M $NaNO_3$. Uranium uptake of the resin (in batch) was 97% from a solution containing 20 ppb uranium, 0.01 M $NaNO_3$ and a pH of 7.

8. Alkaline leaching

Chemistry of carbonate leach

As in the acid leaching, uranium must be in the hexavalent state in order to be recovered by IER from carbonate leach liquors. The most frequently used leaching solution is a mixture of sodium carbonate and sodium bicarbonate, with or without oxidant, usually air. In carnotite ores ($K_2(UO_2)_2(VO_4)_2 \cdot 3H_2O$), uranium is already in the hexavalent state and the reaction is:

$$K_2(UO_2)_2(VO_4)_2 \cdot 3H_2O + 6\ CO_3^{2-} \rightarrow 2\ K^+ + 2\ [UO_2(CO_3)_3]^{4-} + 2\ VO_3^- + 4\ OH^- + H_2O \quad (8.1)$$

From uraninite, UO_2, where uranium is in the tetravalent state, the reaction is:

$$2\ UO_2 + 6\ CO_3^{2-} + 2\ H_2O + O_2 \rightarrow 2\ [UO_2(CO_3)_3]^{4-} + 4\ OH^- \quad (8.2)$$

The stability constant of the uranyl tricarbonate complex is very high, of the order of 10^{22}.

In both cases, OH⁻ are formed which can precipitate uranium as sodium diuranate:

$$2\,[UO_2(CO_3)_3]^{4-} + 6\,OH^- + 2\,Na^+ \rightarrow Na_2U_2O_7 + 6\,CO_3^{2-} + 3H_2O \qquad (8.3)$$

In order to avoid this from happening, bicarbonates are added to neutralize the solution.

The leach liquors have typically the following composition:

> Uranium: 150-500 mg U_3O_8/L
> Na_2CO_3: 10 to 50 g/L
> $NaHCO_3$: 5-10 g/L
> Cl-, SO_4^{2-} : a few g/L
> VO_3^- (carnotite ores) : 300-500 mg V_2O_5/L
> pH about 10

The ion exchange reactions involved are:
$$4\,R\text{-}X + [UO_2(CO_3)_3]^{4-} \rightleftarrows R_4\text{-}[UO_2(CO_3)_3] + 4\,X^-$$
$$2\,R\text{-}X + CO_3^{2-} \rightleftarrows R_2\text{-}CO_3 + 2\,X^- \qquad (8.4)$$
$$R\text{-}X + HCO_3^- \rightleftarrows R\text{-}HCO_3 + X^- \qquad (8.5)$$

where X^- is the ionic form of the resin after elution ($X^- = NO_3^-$, Cl⁻ or HCO_3^-) and R is a SBA resin.

Absorption of uranium

SBA resins are used in uranium recovery from alkaline leach liquors. SX with weak bases and WBA exchange resins are not

suitable here however, liquids with phosphate or phosphonium functional groups have been introduced (Zhu *et al*, 2014b).
The saturation capacity of SBA resins in carbonate leach solutions depends greatly on the total carbonate and bicarbonate concentrations. At a level of 30 g/L Na_2CO_3 and 10 g/L $NaHCO_3$ the saturation capacity of a low moisture gel type resin is of the order of 50 g U_3O_8/L_R or more.

VO_3^- and VO_4^{3-} can be fixed on a SBA resin. Again, the capacity of the resin depends on the total carbonate and bicarbonate concentrations. It has been found that vanadates are fixed significantly better from a HCO_3^- solution than from a CO_3^{2-} solution probably due to the higher resin selectivity for CO_3^{2-} ions (divalent) over HCO_3^- ions (monovalent). Therefore, if both uranium and vanadates are found in the pregnant solution, the absorption of vanadates is minimized if the pH of the solution is higher than 10. Under these conditions and provided that the cycle goes to the saturation of the resin, the vanadium fixed can be less than 1% of the uranium fixed on the resin.

Molybdenum is also fixed by SBA resins but the selectivity of the resin for uranium is higher and molybdenum is displaced by uranium and leaks first. At saturation, the resin contains very little molybdenum so that separation of these two elements is very good.

Sulfates and chlorides can be tolerated in concentrations in the leach solution up to one or two grams per liter. At higher levels they decrease significantly the operating capacity of the resin. For example, chlorides at the level of 4 g/L can decrease the operating capacity for uranium by about 35% while at 2 g/L the decrease is about 15%.

Elution

Sodium chloride and sodium nitrate at molar concentrations containing about 0.1 M Na_2CO_3 have been found to be very efficient eluents. $NaHCO_3$ is frequently used as eluent because it avoids the introduction of Cl^- or NO_3^- in the circuit. Na_2CO_3 was found to be very inefficient eluent for the R_4-$[UO_2(CO_3)_3]$ complex, the same as $(NH_4)_2SO_4$ was found to be inefficient eluent in sulfuric acid leach for the R_4-$[UO_2(SO_4)_3]$ complex (page 167).

The elution reaction is the opposite of the loading reaction, eq. 8.3:

$$R_4\text{-}[UO_2(CO_3)_3] + 4\ X^- \leftrightarrows 4\ R\text{-}X + UO_2^{2+} + 3\ CO_3^{2-}$$

Where $X = Cl^-$, NO_3^- or HCO_3^-.

High temperatures (40-60°C) have a positive effect on the elution efficiency.

After elution and uranium precipitation with NaOH:

$$2\ Na_4UO_2(CO_3)_3 + 6\ NaOH \rightarrow Na_2U_2O_7 + 6\ Na_2CO_3 + 3\ H_2O$$

the barren Na_2CO_3 eluate is recycled back to the leaching section. If both vanadium and uranium are fixed on the resin, Na_2CO_3 can be used first to elute vanadium (with a minimum elution of uranium) following by $NaHCO_3$ elution to recover uranium

Recovery of uranium with carboxylic resins

In order to avoid some drawbacks of the SBA exchange resins, such as chemical degradation of the strong to weak base or fouling, the use of weak acid cation exchange resins in the H^+ form has been proposed (Carmen and Kunin, 1986; Kunin and La Terra, 1986).

The hydrogen form carboxylic groups of the resin decomposes the carbonate complex of uranium and the cationic UO_2^{2+} is then fixed on the resin:

$$2\ RCOOH + UO_2(CO_3)_3^{4-} \rightarrow (RCOO)_2UO_2 + 2\ CO_3^{2-} + H_2CO_3$$

This reaction can not take place if the carboxylic group is in some cationic form, Na^+, Ca^{2+} etc. and therefore the above reaction stops when the resin becomes exhausted with cations. For this reason, two columns in series were suggested in order to increase the cycle length.

Elution of uranium is done with acid which can be reused with acid makeup for a number of times.

Removal of uranium from drinking water

In drinking water in the presence of CO_3^{2-} ions, uranium is found as carbonate complexes, as indicated in the following table:

pH	3	5	7	9
Species	UO_2^{2+}	UO_2CO_3	$UO_2(CO_3)_2^{2-}$	$UO_2(CO_3)_3^{4-}$

Removal of uranium from drinking water follows therefore the same chemistry as in recovering uranium from alkaline leach solutions.
In drinking water, uranium levels vary. Typically they do not exceed the 10 µg/L but in some cases they may be as high as 100 µg/L.

Strong base anion exchange resins are suitable to remove uranium from drinking water. In a study (Hanson *et al*, 1986) it was demonstrated that different commercial SBA resins were able to remove uranium from water from a feed concentration of 200-400 µg/L down to <14 µg/L for 12000 to 14000 BV. Regeneration was done with NaCl. In another study (Wang et al, 2008) SBA resins impregnated with iron were used to remove arsenic (with the iron) and uranium (with the SBA sites) from drinking water. With uranium content in the feed water of about 30 µg/L, the cycle run at 11 BV/h, lasted for more than 30000 BV.

Acrylic and styrenic weak base resins have also used for uranium removal from ground water (Riegel and Höll, 2008). Acrylic gave a beter loading capacity, possibly because acrylic WBA resins are a little stronger bases than styrenic. One important factor, as expected, was the pH of the feed solution. At pH=7-7.5 and a uranium concentration of 1000 µg/L in carbonated water, the acrylic resin (Amberlite™ IRA67 of Rohm and Haas Company) treated more than 30000 BV before breakthrough.

9. Unconventional uranium resources

Uranium recovery from phosphoric acid

One of the processes for production of phosphoric acid is the so called wet process. With this process concentrated (93%) H_2SO_4 is added to the phosphate rock, in the form of fluoroapatite, $3Ca_3(PO_4)_2 \cdot CaF_2$, where the Ca is removed as gypsum by filtration and H_3PO_4 is produced at a concentration 26-32% P_2O_5 or 40-52% P2O5 depending on the precipitation conditions of $CaSO_4$. The concentration of H_3PO_4 is usually expressed as % phosphoric anhydride, P_2O_5, rather than % H_3PO_4.

$$Ca_{10}F_2(PO_4)_6 + 10\ H_2SO_4 + 10\ H_2O \longrightarrow 2\ HF + 6\ H_3PO4 + CaSO4 \cdot 2H_2O$$

Uranium is transferred to a large extent from the phosphate ore to the phosphoric acid. After filtering off gypsum, phosphoric acid is concentrated through evaporators to reach 35% or 54% P_2O_5 or further to produce "superphosphoric acid" of 70% P_2O_5.

The wet phosphoric acid (WPA) thus produced contains impurities such as sulfuric acid, hydrofluoric acid, hydrofluorosilisic acid, Cd, Cr, Fe, Al, V, Ca, Mg, Na or U.

The removal of uranium from WPA has attracted particular interest because not only it purifies H_3PO_4 for certain applications such as food or fertilizers but the uranium removed can be recovered as strategic material for further use.
Phosphate rock can contain uranium from 100 to 200 ppm, depending on the ore. The removal of uranium from phosphoric acid is achieved with solvent extraction, the more usual being the DEPA-TOPO solvent extraction (DEPA=D2EHPA= di-2-ethyl hexylphosphoric acid and TOPO=tri-octylphosphine oxide). However, ion exchange can be used using chelating resin of the aminomethylphosphonic acid type or with inert resins impregnated with di-2-ethyl hexylphosphoric acid (D2EHPA) (Gonzalles-Luque and Streat, 1984; Volkman, 19867; Hassid *et al*, 1986; Kabay *et al*, 1998; Soldenhoff *et al*, 2009).

Before passing through the AMP resin, uranium is reduced with metallic iron powder. With this reduction, uranium is converted to the U(IV) oxidation state. AMP resins can remove both U(IV) and U(VI) but U(IV) can be removed from a more concentrated H_3PO_4 than U(VI) (Soldenhoff *et al*, 2009). At the same time, Fe(III) is reduced to Fe(II) and the reduction of iron is an important feature of the process because Fe(III) is also fixed on the AMP resins but it poisons the AMP resins (Volkman, 1987). In addition, U^{4+} is fixed stronger than Fe^{2+} so that U^{4+} displaces Fe^{2+} from the resin.
In another process (Bristow *et al*, 2013) the iron content of the WPA is previously reduced by precipitation as iron (III) ammo

nium phosphate, following by reduction of the remaining Fe(III) to Fe(II).

The fixation of uranium by the resin from WPA is a slow process. The operating capacity of the resin depends greatly on the specific flow rate (BV/h) and on the resin particle size. Operating capacities of 6-10 g U/L_R can be obtained.

Elution of the uranium from the resin is achieved with a mixture of NH_4HCO_3 and $(NH_4)_2CO_3$. Before elution, the uranium on the resin is oxidized with raw (unreduced) phosphoric acid. The purpose of this is to remove Fe^{2+} loaded along with U on the resin so that U is recovered in purer condition. Then, the resin is washed with NH_4OH in order to neutralize the sites in H^+ form to avoid the formation of CO_2 during the $(NH_4)_2CO_3$ elution, and rinsed.

Uranium from seawater

Uranium is found in the seawater as the uranyl tricarbonate complex, $[UO_2(CO_3)_3]^{4-}$, at a level of 3.3 ppb. The total uranium in the oceans is 4 billion tons. In view of the fact that a 1-gigawatt nuclear power plant needs 27 tons of uranium fuel per year, there is enough uranium in the ocean to cover the energy needs of the planet for about 10 000 years.

In the years 1960's, it was found that the best way to extract uranium from seawater was hydrous titanium dioxide, TiO_2. Later, in the 1980's, it was determined that amidoxime functional groups were more efficient and more stable than TiO_2. Uranium forms coordination bonds with two amidoxime groups:

$$-R-C\begin{array}{c}\nwarrow NOH\\\diagdown NH_2\end{array} \quad \begin{array}{c}H_2N\diagdown\\HON\diagup\end{array}C-R- \; + \; [UO_2(CO_3)_3]^{4-} \longrightarrow$$

$$-R-C\begin{array}{c}\nwarrow NO\\\diagdown NH_2\end{array}UO_2^{2+}\begin{array}{c}H_2N\diagdown\\ON\diagup\end{array}C-R- \; + \; 3\,CO_3^{2-} \; + \; 2\,H^+$$

Adsorbents were then developed by grafting acrylonitrile units on polyethelene fibers, and then converting to amidoxime functional groups with hydroxylamine at an appropriate pH and temperature:

$$-(CH_2-\underset{\underset{C\equiv N}{|}}{CH})_n- \; + \; NH_2OH \; \longrightarrow \; -(CH_2-\underset{\underset{\underset{NH_2}{|}}{C=N-OH}}{CH})_n-$$

The operating capacity of this kind of fibers is of the order of 6 g U/kg of adsorbent in a time of 8 weeks.

More recently, Oak Ridge National Laboratory (ORNL) has developed a new amidoxime-based adsorbent by electron beam induced grafting of acrylonitrile (AN) and itaconic acid (ITA) on polyethylene hollow fibers (Tsouris *et al*, 2015). AN is subsequent converted to amidoxime groups with hydroxylamine. One key factor that affects uranium uptake is the alkaline treatment of this material (Tsouris et al, 2015). This treatment has two effects: converts the carboxylic groups of the ITA to the carboxylate form thus making the material more hydrophilic and second, converts the amidoxime groups to carboxylate and also the open-chain amidoxime groups to the cyclic imidedioxime:

$$\begin{array}{c} \text{NH}_2 \quad \text{NH}_2 \\ | \quad\quad | \\ \text{HO-N=C} \quad \text{C=N-OH} \\ | \quad\quad | \\ \text{-C-C-C-C-} \end{array} \longrightarrow \begin{array}{c} \text{NH} \\ / \;\; \backslash \\ \text{HO-N=C} \quad \text{C=N-OH} \\ | \quad\quad | \\ \text{-C-C-C-C-} \end{array} + \text{NH}_3$$

Uranium forms stronger complexes with two cyclic imidedioxime units than with two adjacent amidoxime units (Britt et al, 2014):

[Structural diagram of UO_2^{2+} complex with two cyclic imidedioxime units]

Elution of uranium from amidoxime fibers can be done with 0.5-1 M HCl or with a mixture of 1M Na_2CO_3+ 0.1 M H_2O_2. If HCl elution is not used, the alkali treatment is not necessary. HCl elution may decrease the amidoxime capacity of the material.

10. Ion exchange systems in uranium recovery

Fixed bed systems, three columns in series

In fixed bed systems, resins with an average particle size of 0.6 to 0.7 mm are used. The flow rates employed are determined in such a way that in a three column merry-go-round system with two columns on loading and one on elution, the first loading column will be close to saturation while the second will start to break through. They are usually in the range of 5 BV/h where BV means the total resin volume of the two loading columns.

The processing of the pregnant liquor comprises the following steps :

- absorption cycle
- rinse
- backwash and settling
- elution
- backwash and settling
- resin sulfatation or carbonatation
- standby

Absorption cycle

The ion exchange reactions in the adsorption cycle are those given in (7.6) to (7.9) for acid leach and in (8.3) to (8.5) for alkaline leach. In both cases, X^- is the ionic form of the resin after elution. The leakage profiles from the first and second column are illustrated in the following figure :

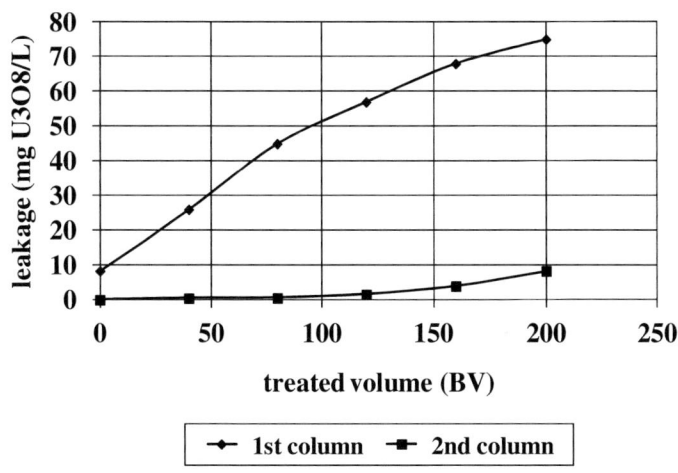

Figure 9.1 Leakage representation of a merry-go-round system

The cycle length can be calculated from the uranium concentration in the pregnant solution, the specific flow rate (BV/h) and the saturation capacity of the resin. *For the cycle length calculation, one BV is the resin volume in only one column.* For example, with a pregnant solution containing 200 mg U_3O_8/L at 8 BV/h and with a saturation capacity of the resin of 48 g U_3O_8/L_R, the cycle time will be 30 hours ((48*1000/200)/8).

Rinse

When the second column starts breaking through, the feed flow stops and the columns are rinsed with water from the first into the second column at the same flow rate as the absorption cycle. The water rinse displaces the pregnant solution still in the resin and the interstitial volume. In general, 1-3 BV is enough.

Backwash and settling

After rinse, the first column is isolated while the second is ready for the next absorption cycle. The third column which has been on standby is placed after the second column.
The first column undergoes a backwash step. The purpose of the backwash operation is to remove any suspended solids that have accumulated on top of the resin bed and which can cause channeling and high pressure drop across the resin bed. To eliminate these suspended solids and eventually any small size broken resin, the resin bed is fluidized by passing water upflow at a linear velocity such that the resin remains in the column but the suspended matter are removed in the overflow. An efficient backwash requires at least 50% resin expansion. Fluidization curves giving the resin expansion as a function of linear velocity for SBA resins in the exhausted form are generally given by the resin manufacturers. Approximately, 4-5 BV at velocities in the range 10-15 m/h are needed. The effluent of the backwash is returned to the filters. After backwash, the resin is allowed to settle before the elution step.

Elution

As seen in figure 7.4 page 167, most of the eluted uranium comes off the resin during about the first half of the eluent passed through the resin. In order to save eluent it is possible to recycle part of the excess to the following elution, as follows : the first half of the eluent comes from the recycled eluent (second half of the previous elution) while the second half is fresh eluent which will be used for the first half of the following elution. Thus, the first part is the concentrated eluate and the second half the eluate to be recycled.

In the case of uranium recovery using conventional fixed bed column system, eluent recycling can be done in the following way :

stage	from	to
(a)	recycle	pregnant
(b)	recycle	conc. eluate
(c)	fresh	conc. eluate
(d)	fresh	recycle
(e)	water	recycle

In the case of NO_3^- or Cl^- elution in acid leach, uranium is precipitated from the concentrated eluate by ammonia, caustic, magnesia or hydrogen peroxide. The barren eluate however contains HSO_4^- and SO_4^{2-} ions coming from the uranyl sulfate complex, the eluent as well as from the ionic sites of the resin in the SO_4^{2-} or HSO_4^- form. In order to prevent SO_4^{2-} build up some plants use two stage precipitation. In the first stage, lime is added to a pH of about 3.5 where SO_4^{2-} are removed as $CaSO_4$ but also most of the iron is precipitated thus resulting in a purer uranium. This precipitate, called iron cake, returns to the leaching. Then the pH is raised to about 7.0 where uranium precipitates. The barren eluate is then recycled as fresh eluent. In the case of H_2SO_4 elution, uranium is recovered by solvent extraction and the barren eluate is again recycled as fresh eluent.

In the case of $NaHCO_3$ elution in alkaline leaching, uranium is precipitated with sodium hydroxide and the barren eluate is recycled to the leaching.

For mass balance purposes, the volume of stage (a) should be equal to that of stage (e) and the volume of stage (b) should be equal to that of stage (d).

Backwash and settling

The purpose of the final backwash is to remove the remaining eluent from the resin or from the upper part of the column. This avoids an initial small uranium leakage during the beginning of the absorption cycle caused by the eluent as it is displaced by the incoming feed solution. Since at this point the resin is found in the ionic form of the eluent, the linear velocities for backwash should be lower than in the previous backwash step where the resin was exhausted and therefore heavier.

Resin sulfatation or carbonatation

In case of Cl^- or NO_3^- elutions, in order to avoid introduction of Cl^- or NO_3^- ions in the leach liquor, the resin can be converted back to the ionic form it has in the leaching solution, SO_4^{2-} for acid leach and CO_3^{2-} for alkaline leach, before the absorption cycle. For that, it can be washed with H_2SO_4 in the case of acid leaching or with Na_2CO_3 in the case of alkaline leaching and the effluent is recycled to the eluent.

In case where weak base resins are used to recover uranium from acid leach PLS, this step is done easier by an alkaline or ammoniacal wash at a stoichiometric level to convert the resin to the free base form. The eluate is recycled to be reused as eluent. The resin is subsequently protonated by the PLS. care must be taken if silica is found in the resin and is transferred to the eluate.

Continuous ion exchange

One important step in the flowsheet of treating the ores to recover uranium is the solids-liquid separation which includes the leached ores washing and solution clarification. When fixed bed ion exchange is used, a clarified liquor is necessary. In some cases the leach liquors exhibit poor filtering and settling characteristics and necessitate very large size liquid-solids separation equipment. Systems that allow treating liquors containing high content of suspended solids and even pulps offer a significant cost reduction. These systems combining continuous ion exchange (CIX) allow significant cost reductions and are used widely in uranium hydrometallurgy. Some of them are described below.

Among the continuous ion exchange systems developed for uranium recovery, those mainly used today are: Fluidized bed (the NIMCIX and the Porter designs, in S.Africa and Namibia), moving packed bed (ex Soviet Union countries) and Resin-In-Pulp (RIP).

Fluidized bed

In the fluidized bed systems, the pregnant solution flows in upflow direction through a tower or other type of container, divided into separated stages in order to avoid resin mixing, thus causing a fluidization of the resin. In this way, the suspended solids in the liquor can go out from the top without clogging the resin bed. Given a pregnant liquor flow rate and the uranium concentration in it, the question then becomes how many stages and how much resin per stage are needed to ensure a given uranium concentration in the barrens. The independent variables that need to be fixed beforehand are the degree of fluidization of the resin, the settled resin height and the stage height, making sure that the fluidized resin height is less than the stage height.
The NIMCIX system was developed in South Africa in the years 70's by the National Institute for Metallurgy (NIM, today Mintek) based on the Cloete-Streat concept (Cloete and Streat, 1967 ; Cloete and Haines, 1971 ; Haines, 1972; NIM, 1978) and it is still in use today. The principle is outlined in figure 9.2.
The system consists of an absorption column (AC), a measuring column (MC), and an elution column (EC). In addition, there is a regeneration column (RC) where the resin is periodically treated to eliminate silica that may foul the resin. The absorption column is essentially a cylindrical tower divided into a number of stages (in the figure there are 14 stages) separated by perforated trays. The openings of the trays are bigger than the largest size of the resin beads. Each stage contains a given quantity of resin. In principle, each stage contains the same quantity of resin in practice however, this quantity can vary. The feed solution is pumped upflow and the resin in each stage is fluidized. After a

certain period of time, the flow of the liquid stops and the resin is allowed to settle to the bottom of each stage. Then the flow is reversed and during this period, a given volume of resin from the bottom stage moves into the measuring column and at the same time, resin is transferred from each stage to the stage below. Then resin of equal volume as the resin in the MC is transferred under hydraulic pressure and using barren solution, from the top of the elution column to the top stage of the AC. The resin from the MC is then transferred hydraulically to the bottom of the EC. The EC contains a number of resin slugs and the column can be a packed bed column or a fluidized column similar to the AC. At every resin transfer, the resin in the EC moves one slug up. Eluent flows in a counter-current way from the resin so that the resin transferred to the AC has seen fresh eluent last.

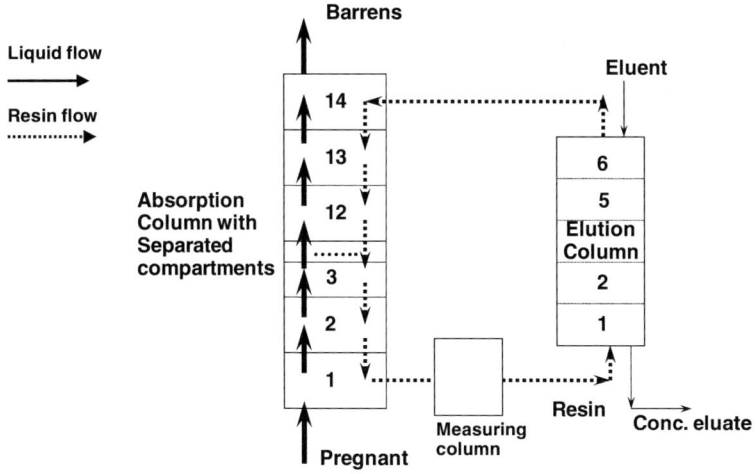

Figure 9.2 NIMCIX system

Typical values for this system are 100% degree of resin fluidization in the AC and one meter stage height, therefore the settled resin height is 50 cm. The dimensions of the absorption column are determined by the hydraulic behavior of the resin. For example, assume that the feed solution has a flow rate of 200 m3/h. In order to have an expansion of 100% of the resin, the linear velocity of the solution needs to be (from the fluidization curves of the resin in the corresponding ionic form) about 15 m/h. This results in a cross-sectional area of the column of 200 / 15 = 13.3 m^2, or a diameter of 4.1 m. In fact, since the density of the resin along the stages varies, the higher density being that of the resin saturated with uranium, the linear velocity taken into account is that needed to fluidize the resin by 100% in the SO_4^{2-} form which corresponds to the top stage. The stage volume will be 13.3 m^3 and the resin inventory per stage 6.65 m^3. In the case of the example of figure 9.2 where there are 14 stages, the column height will be 14 meters (approximately) and the total resin inventory in the AC should be 93 m^3.

The number of stages necessary to bring the uranium concentration from the value in the pregnant solution down to the desired barrens value can be calculated based on the McCabe-Thiele approach which assumes that equilibrium is reached at every stage. Alternatively, it can be calculated from equilibrium isotherms and the rate of uranium loading on the resin to reach equilibrium (Wright 1979). The calculation of the number of stages needed goes as follows. One should calculate the uranium concentration in solution in each stage that comes to equilibrium with the resin, and the uranium loading on the equilibrated resin. The starting point is the top stage (number 14 of figure 9.2) and it is given the desired solution concentration at equilibrium exiting the AC (which is the barrens concentration) and the uranium loaded on the resin at the beginning of the cycle (which is the residual uranium on the eluted resin). From the barrens concentration one can calculate the corresponding equilibrium resin loading. The difference between the initial resin loading (the residual uranium) and the equilibrium loading at the end of the cycle time (in equilibrium with the barrens) should come from the solution flowing from stage 13 to stage 14 during the cycle time. One therefore can calculate this solution concentration. In fact, since the liquor flows continuously, the concentration of the liquor coming out of each contactor taken into account is the average concentration over the cycle time. This solution concentration in turn is used to calculate the resin loading in equilibrium. The difference between this equilibrium value and the equilibrium loading of the resin at the end of the cycle time in stage 13 should come from the solution flowing from stage 12 to stage 13. One can again calculate this solution concentration. The calculations continue in the same way until the solution concentration reaches the value of the pregnant solution corresponding to the bottom stage (number 1 in fig. 9.2). In fact, one can also introduce a rate constant which allows the calculation of the extent to which the resin loading reaches the equilibrium at each stage within the cycle time. This rate constant can be obtained experimentally as discussed in the last Chapter. For these calculations, the parameters used are the quantity of resin transferred from one stage to the previous one and from the AC to the EC and back, and the time between resin transfers (cycle time), which also defines the resin flow rate. This calculation works when the cycle time has been calculated by dividing the amount of uranium that the resin going to the elution column can load by the rate that uranium enters the AC. For example, with a feed flow rate of 270 m3/h and a uranium concentration of 0.13 kg U/m3

One of the earliest fluidized bed installations in South Africa using the NIMCIX system was the Chemwes uranium plant, commissioned in 1979 (McIntosh et al, 1982). This plant contained two parallel lines, each with NIMCIX absorption and elution columns. Each AC had a diameter of 4.85 m and treated 6500 m3 PLS/day, or 271 m3/h. the resulting superficial velocity of 14.7 m/h corresponds to a 100 % fluidization of a typical gel type SBA resin of an average particle size about 0.65 mm. The AC contained 12 compartments, calculated as described above, in such a way that the barrens were < 2 ppm and considering a resin capacity of 25 g U_3O_8/L_R. The cycle time can be adjusted based on the uranium concentration in the feed solution.

The EC will be dimensioned based on the resin flow rate and the elution profile of the resin (eluent specific flow rate and eluent volume (BV) required to elute uranium down to a given residual value, for example <1 g U_3O_8/L_R). If for example a residence time for elution required is 12 hours and the specific flow rate is 1 BV/h, then the eluent volume needed is 12 BV. If every 3 hours 4 m3 of resin is transferred from the AC to the EC, then there will be 12/3 = 4 stages in the EC each with 4 m3 resin, or a total of 16 m3 of resin. The eluent volume needed is 12 BV * 16 m3R = 192 m3 while the eluent flow rate is 192/12 = 16 m3/h. With a resin capacity 40 g U_3O_8/L_R, while the stripped resin contains 1 g U_3O_8/L_R residual uranium, then 39*4/3 = 52 kg U_3O_8 will be eluted every hour. The concentrated eluate concentration will then be 52/16 = 3.25 g U_3O_8/L.

In the Porter design (Porter, 1971) the absorption column is in fact a series of horizontal tanks each tank acting as a fluidized bed. The resin is transferred from one tank to the next one by airlifts. The pregnant solution is pumped from one tank to the previous one in counter-current way to the resin. Elution is done in three packed bed columns in series. The resin is transferred to the first column while eluent comes in from the last column towards the first. This system is in operation in one of the biggest uranium mines in the world, Rossing Uranium in Namibia. The contactors are 6.2 meters long by 6.2 meters wide and contain about 30 m^3 resin per contactor. The flow rate per line being about 800-850 m^3/h, linear velocity of the liquor flowing through is about 22 m/h. There are five contactors with a sixth acting as resin trap. There are three elution fixed bed columns in series per line containing each 22 m^3 of resin. If the saturated resin goes to column number 1 then the eluent, H_2SO_4, flows from column number 3 to column number 1, counter-current to the resin.

Moving packed bed

Figure 9.3 illustrates the principle of the moving packed bed continuous system, practiced in the former Soviet Union countries. The system consists of an absorption column, a transfer column and a series of elution columns, usually 3-4. The resin in the transfer column is washed to remove the suspended solids that have accumulated in the resin during the loading stage. In addition, before the elution columns there is a scrubbing or saturation column through which passes part of the concentrated eluate. After the elution columns there are 1 to 2 columns to resulfate the resins when nitrate elution is practiced, and one column for a final water rinse (Patrin, 2005).

The resin fills completely the absorption column and is maintained inside the column by screens placed at the top of the column. The screen size determines the particle size of the resin to be used. The specific flow rate is typically 5-7 BV/h. The pregnant solution is fed from the bottom to the top. The size of the AC is determined by the volumetric flow rate and the specific flow rate. For example, for a pregnant solution flow rate of 200 m^3/h and a specific flow rate of 5 BV/h, the resin volume is 40 m^3. With a vessel of 3 meters diameter (7 m^2 cross-sectional area) the resin bed height and hence the column height, is 5.7 meters. The linear velocity of the liquid is approximately 30 m/h.

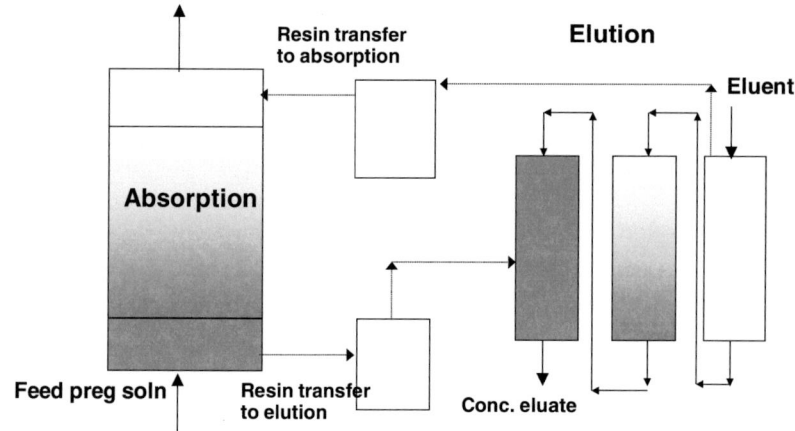

Figure 9.3 Principle of moving packed bed system

Because of the tall resin bed and the fast linear velocity, pressure drop may be high, especially since suspended solids are accumulated at the bottom of the column. For that reason, in many mines using this technology they use resins having bead diameters bigger than 0.8 mm. The screens at the top in that case are 0.6 mm.

Every certain time intervals, typically every 3-4 hours, about 15% of resin volume is transferred from the bottom part of the column into the transfer vessel where it is cleaned from the accumulated suspended matter by washing over screens where the suspended matter is removed and the clean resin is transferred to the elution. The elution section can be a series of fixed bed columns. Fresh eluent is fed from the last towards the first elution column while the resin is transferred from the absorption column and goes to the first elution column. Before elution the resin undergoes a scrubbing step (or an over-saturation step) where part of the concentrated eluate goes to the resin that has just been transferred from the AC and then recycled to the pregnant solution tank. During this scrubbing step, the uranium loading on the resin increases (except in the case of H_2SO_4 elution) and this contributes to achieve a more concentrated and purer eluate. With the above system and using a 70-80 g NO_3/L eluent containing 25-30 g/L H_2SO_4, the concentrated eluate contains around 25-35 g U/L and 20-25 excess acid. After elution, the resin is transferred to a transfer vessel where it is converted back to the leach solution form. For example, if NO_3^- elution is practiced, the resin is washed with H_2SO_4 to convert it to the SO_4^{2-}/HSO_4^- form. The spent H_2SO_4 is then used to acidify the NH_4NO_3 eluent.

Another type of elution column is a U-shaped column, called Sorption-Desorption-Contour (СДК) column used in certain mines in Kazakhstan (Kudlanov, 2000; Patrin, 2005) in replacement of the columns in series shown in figure 9.3 above. As the name implies, the column has a U shape and it is outlined in figure 9.4.

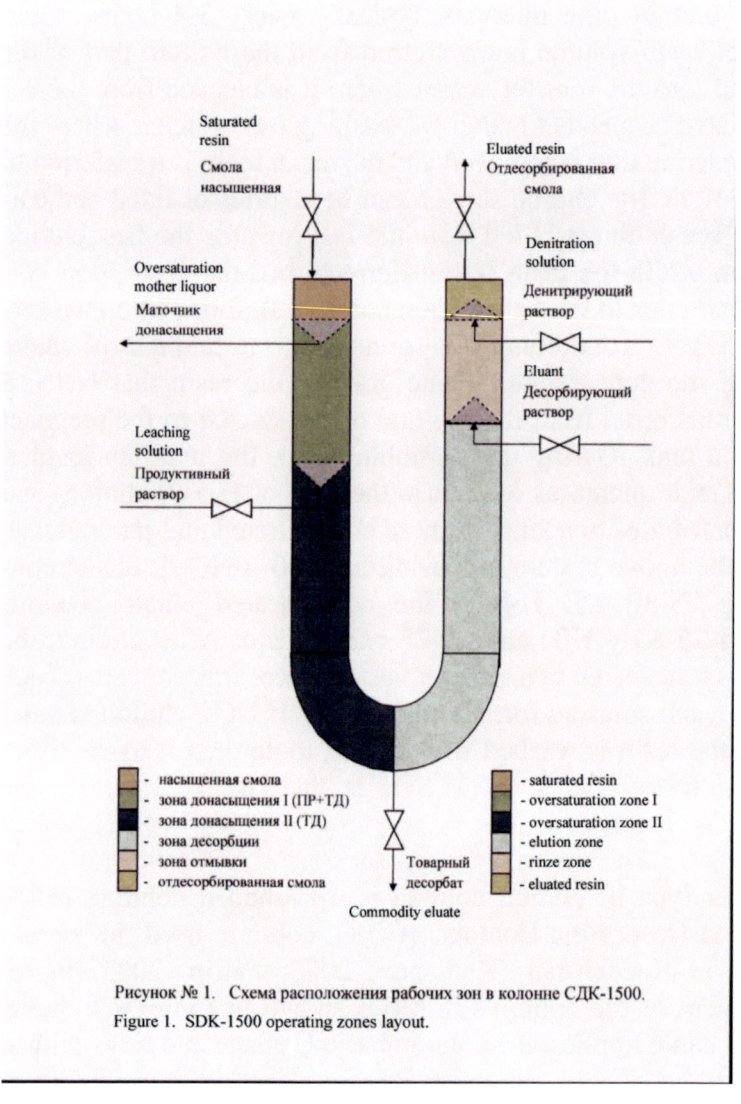

Figure 9.4 The СДК-1500, U-shaped elution column (Courtesy of NAC Kazatomprom)

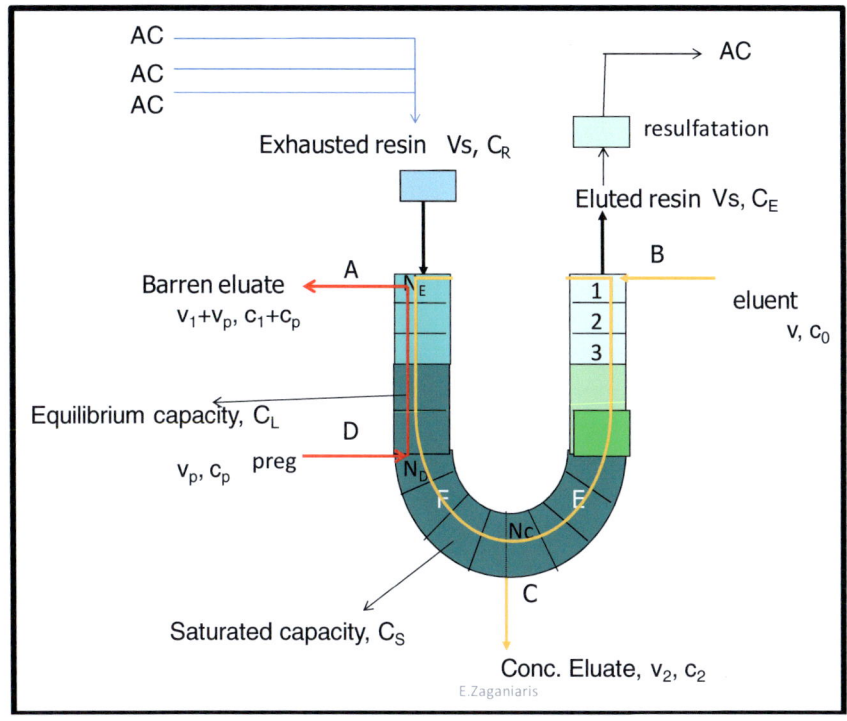

Fig 9.4a U-shaped elution column outline

The saturated resin is transferred from the absorption column and enters the elution column from the left side indicated in figure 9.4. At every resin transfer it is pushed through until it reaches the opposite end. The eluent enters the column from the other end as shown in the diagram, in counter current mode. The last part of the column is used to resulfate the resin, when nitrate elution is practiced. The concentrated eluate is split in two parts, one part leaves the column from the bottom of the U column and goes to the uranium precipitation and recovery while the other part goes through the exhausted resin and leaves the U column from the side the loaded resin is introduced (fig. 9.4a). By doing this, the saturated resin still found in the left side of the column picks up more uranium from the concentrated eluate and becomes over-saturated and at the same time impurity ions such as $Fe(SO_4)_2^-$ and SO_4^{2-} go into the out-going barren eluate which returns to the pregnant solution. The ratio of the two streams of the split eluate will determine the degree of over-saturation of the resin and the uranium concentration in the concentrated eluate part.

The U-shaped column is reported (Patrin, 2005) to bring the following advantages with respect to the conventional separate elution columns (fig. 9.3): improved the overall kinetics of the elution process, decreased elution time, use of lower acidity in the nitrate eluent solution thus improving the corrosion resistance of the installation, obtain higher uranium concentration in the eluate, up to 80-90 g U/L, obtain higher uranium recovery by decreasing the residual uranium on the eluted resin and obtain a better quality of the precipitated uranium.

A U-shaped "desorption/concentration" elution column has also been promoted by Clean TeQ Ltd under the name of Clean-iX® Elution process, using the principle of concentration-desorption (Zontov, 2006; Carr, 2008).

The Clean TeQ U Column (figure 9.5) is a specifically designed desorption concentrator column that is used to desorb a selected metal species from the resin while rejecting impurities.

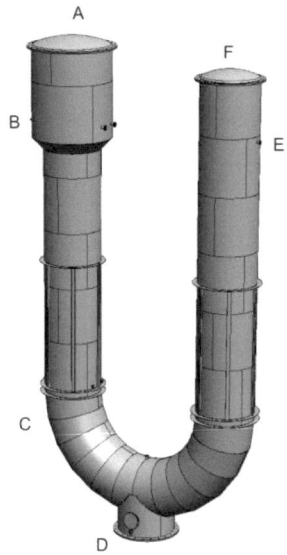

Figure 9.5 Clean TeQ Clean-iX® U column (Courtesy of CleanTeQ Ltd)

The column is U shaped in configuration with resin entering the top left side of the U column (A) and desorption solution entering the opposite side (F). The resin moves counter currently against the desorption solution. Fully desorbed resin is removed from the opposing side (E) of the U column where the loaded resin is added. There are two solution streams removed from the column:
 (i) a pregnant solution containing the targeted metal species in a concentrated form pumped from the base (D) of the U column, and
 (ii) a waste stream which contains the majority of impurities and a small amount of the target metal species which

overflows into a launder from the top of the U column (B).

The key advantage of this type of column design is the ability to concentrate the metal ion of interest in the pregnant solution stream, while at the same time increasing the relative purity of the pregnant solution. This is achieved by an internal desorption and adsorption process within the U column. Where the desorption solution enters the U column (F) the metal ion of interest is desorbed from the resin. As desorption takes place a concentration gradient is formed between the metal species loaded on to the resin and the concentration of metal species in solution. The metal species of interest ideally reaches its highest concentration at the bottom of the U column (between C and D). This increased liquid phase concentration of the targeted metal provides the potential to "scrub" impurities off the resin and fully load the resin with the targeted metal. The relative flows of desorption solution, pregnant solution, waste solution and resin are adjusted to ensure that the concentration gradient for targeted metal remains at its highest near the base of the U column with the waste solution chromatographically pushed to the top of the column.

The effect of the U column is to allow a high concentration, high purity targeted metal to be produced in a single column. Compared with conventional batch desorption processes, the metal concentration of the pregnant solution from the U column can be up to four times more concentrated than the concentration of the equivalent batch process. Additionally, the amount of impurities in the pregnant solution from this process can be significantly less than conventional batch processes.

Resin-In-Pulp (RIP)

The principle of the RIP system is illustrated in figure 5.8 page 110. It consists of a series of absorbing contactors and an elution section. Resin and pulp move counter-currently and separated by screens. Although the chemistry of uranium recovery remains the same as in the fixed bed or the previous CIX systems, the equipment is quite different some of which are outlined below.

The various RIP systems developed differ in the way the resin and pulp are mixed in the contactors and in the way resin and pulp are separated from each other and transported from one contactor to the next one. As a result, different systems can handle different solids content in the pulp and different resin to pulp ratios.

Mixing is achieved by air agitation or by mechanical stirring. Air agitation is more gentle and in general results into lower resin attrition. The resin and the pulp are continuously transported and separated in a counter-current mode. The transport of the resin and the pulp from one contactor to another is done with airlifts or by mechanical pumping which may result into higher resin attrition. The separation of the resin from the pulp is done using screening devices. When carousel arrangement is used, as in the case of the AAC pumpcell circuit described below, it is the pulp feed and the tailings discharge positions that are rotated, instead of the resin being transferred from one contactor to another.

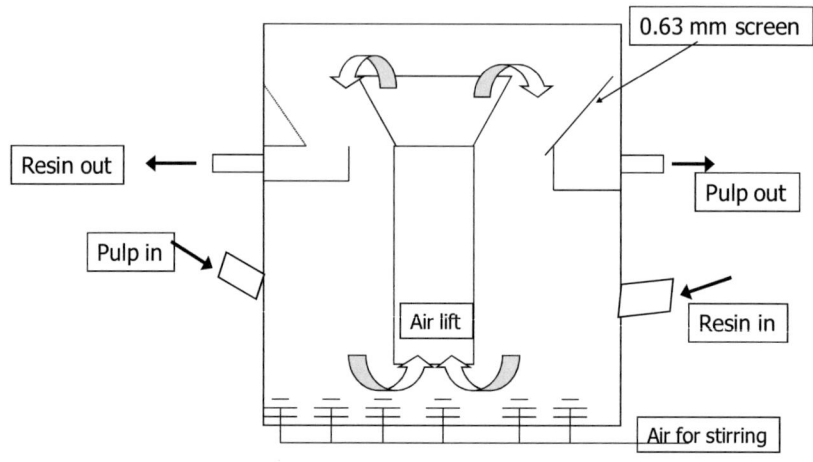

Figure 9.6 Air agitated RIP contactor

Figure 9.6 outlines an RIP contactor. The resin and the pulp are mixed continuously using air agitation. Air lifts are lifting resin and pulp through the inside of a tube and transferred through screens.

One major parameter in separating resin and pulp is the particle size of the resin and the pulp and the openings of the screen. In certain systems the pulp flows through the screens from one contactor to the next one by gravity. In this case, in order that the separation of the resin from the pulp takes place in a reasonably short time, the aperture of the screens should be large compared to the pulp size and the resin particle size should be large compared to the screen openings. For example, the resins for such RIP systems have usually particle size between 0.8 and 1.2 mm. As a consequence of the large resin particle size, the kinetics of uranium loading and elution are slow, as discussed in Chapter 7, page 156. To improve kinetics, some resins for RIP systems have somewhat higher moisture, thus improving the diffusion through the resin matrix. Because of the slow kinetics of the RIP resins, the specific flow rate is generally slow. Another concern is the physical attrition of the resin due to the agitation in the presence of abrasive suspended solids and due to the transfer from one contactor to the other through screens. In some of the existing RIP systems, resin attrition constitutes a major parameter in the economics.

Clean TeQ has developed the Clean-iX® continuous RIP process (cRIP) that can extract uranium from pulps containing up to 50-55% w/w solids and using a resin content up to 40% v/v. The Clean TeQ Clean-iX® cRIP system typically consists of a number of air-stirred reactors in series (generally six to ten). The reactors are arranged such that the resin and pulp flow in opposite directions (counter-current flow) to one another. The number of reactors is determined by the number of transfer steps required to reduce the targeted metal to a nominated residual value and the mixing efficiency of each reactor. The design of the reactor with the integrated air mixer and pulp-resin separation system is critical to the operational efficiency of the process. Figure 9.7 shows a single cRIP contactor.

Within each reactor, the pulp is intimately mixed with the resin using a dedicated air mixing system. Additionally, pulp and resin is continually air-lifted to the upper section of the reactor and passed across separation screens. The aperture of the screen is such that slurry passes through the screen and advances to the next downstream reactor, while the resin remains on the screen and is advanced to the upstream reactor by gravity in a counter-current flow mode. A portion of the resin can be recycled back to the originating reactor if the residence times between the pulp and resin are different. As the pulp moves between reactors, the targeted metal ion is adsorbed from the liquid phase of the pulp onto the resin. After passing through the cRIP process, the barren pulp emerges from the last reactor and is passed across a safety screen to recover any resin which has bypassed the last reactor.

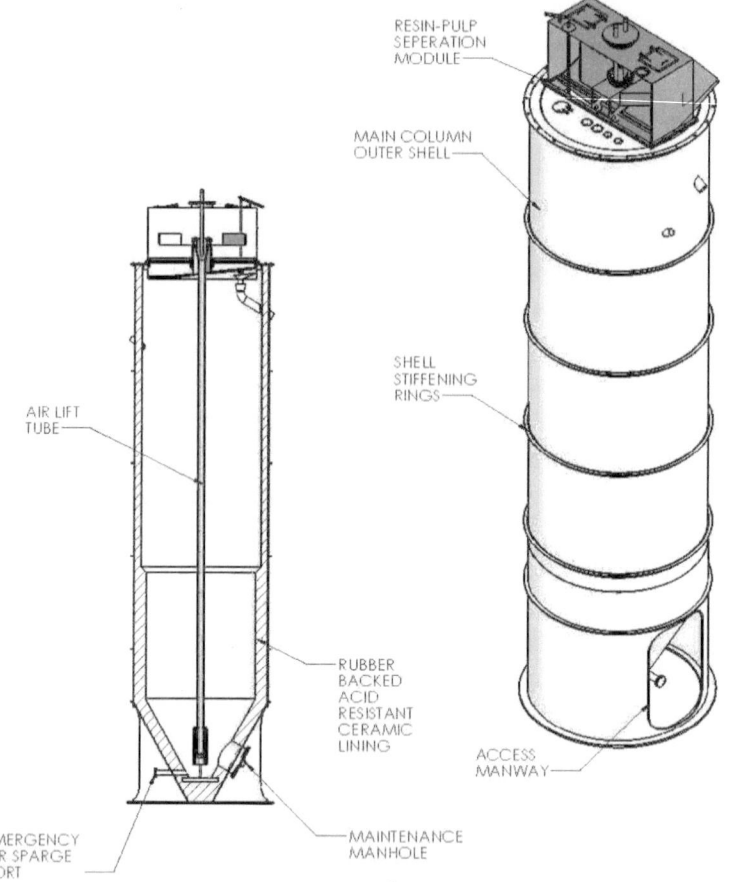

Figure 9.7 Clean TeQ Clean-iX® cRIP contactor (Courtesy of CleanTeQ Ltd)

The pregnant resin emerges from the first cRIP reactor and is transported to the Clean-iX® Elution plant for metal recovery.

Bateman Minerals and Metals and Mintek, both of South Africa, developed jointly a system named MetRIX^{TM} for base metals recovery (van Hege *et al*, 2006) but also suitable for uranium recovery.

The MetRIX^{TM} system consists of a number of absorbing contactors, or stages, in cascade arrangement, two buffer tanks and an elution column. The resin and the pulp are transferred independently. Pulp is pumped into the first stage, then it gravitates to the next stages and finally into the barrens tank. Resin is transferred hydraulically and counter-currently to the pulp. Separation of resin from the pulp is achieved inside the stages through a cylindrical screen.

The Anglo American Corporation (AAC) Pumpcell, licensed to Kemix® is a circuit used in Carbon-In-Pulp (CIP) and RIP applications. Pumping, screening and agitation are combined in a single drive unit, contained in a contactor known as AAC Pumpcell (fig. 9.8).

The AAC Pumpcell circuit has all contactors positioned at the same level. This allows the system to operate in a carousel mode, that is, the pulp feed and tailings discharge positions are rotated to cause a counter-current movement to the resin without the need to move resin from one contactor to another, thus minimizing resin attrition and back mixing. High grade pulp is directed to the designated head contactor. Exiting this contactor, pulp is directed to the next one and so on until it exits the tail contactor to the residue section.

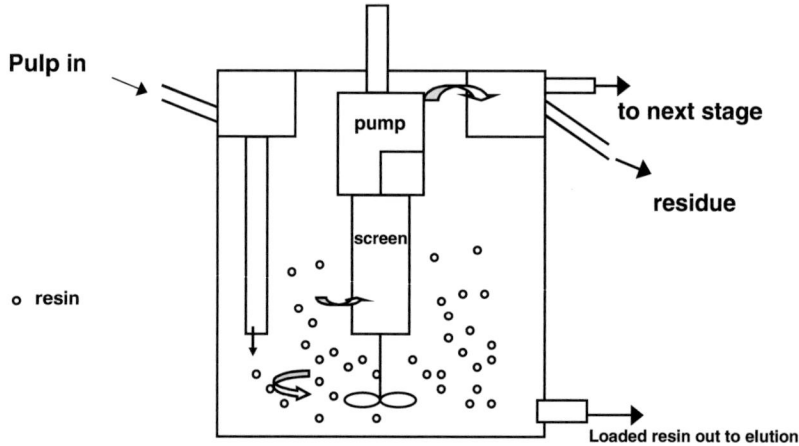

Figure 9.8 Principle of the AAC Pumpcell RIP contactor

Once the resin in the head contactor has reached saturation, this contactor is isolated and the high grade pulp is directed to the next contactor. The contents of the original head contactor are drained and screened to recover the loaded resin.

Then the original head is placed as tail contactor. Freshly eluted resin is added to the new tail contactor while being filled with pulp coming from the original tail contactor. Pulp and resin separation is achieved in each contactor through interstage screen including a pumping impeller. The pumping action generated by the interstage screen allows all contactors to be arranged on the same elevation as the pulp is pumped through the entire Pumpcell circuit.

The Pumpcell mechanism is shown in figure 9.9. Figure 9.10 presents a view of the Pumpcell circuit where the first contactor is seen in more detail.

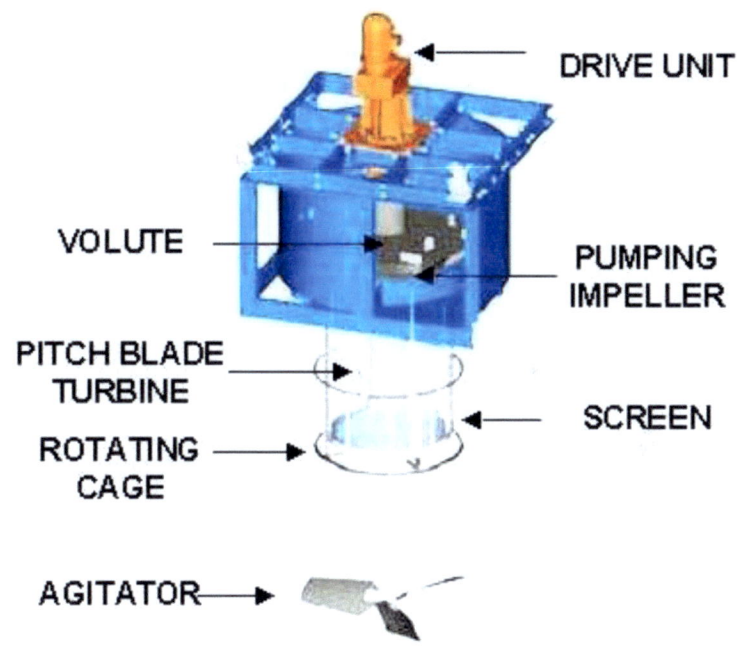

Figure 9.9 Pumpcell mechanism (Courtesy of Kemix (Pty) Ltd)

Figure 9.10 Pumpcell circuit (Courtesy of Kemix (Pty) Ltd)

Flow rates used in RIP systems are rather slow. It depends on the resin and pulp mixing efficiency and on the resin particle size. In general it is in the range of 3-4 BV/h (or a residence time of 20-15 minutes respectively). Contrary to the previous systems, here, by BV it is meant the contactor volume and not the resin volume. This was also the case in the calculations of the resin profiles in batch operations, in pages 108-109.

In dimensioning an RIP installation, another parameter used is the pulp to resin ratio. For example, for a Kemix® AAC Pumpcell system, a pulp to resin ratio of 6.7 (15% resin) and a residence time of 15 to 20 minutes are frequently used.

As an example, using the Kemix® AAC Pumpcell system, with 150 m3/h pulp flow rate and a residence time of 16 minutes (3.75 BV/h) the contactor size will be 40 m3 (= 150 / 3.75). With pulp to resin ratio of 6.7 (15% resin) the resin volume in each contactor would be 6 m3 (=40 / 6.7). Converting the pulp flow rate to clear solution flow rate using the % solids in the pulp and the pulp density, the cycle time is calculated as follows : at 35% solids and 1.283 pulp density, the clear solution flow rate becomes approx. 120 m3/h. At a uranium value of 176 mg U_3O_8/L in the clear solution, every hour 21.1 kg U_3O_8 (=120*0.176) pass through the head contactor. At a resin capacity of 28 g U_3O_8/L_R, the 6 m3 resin can absorb 28*6= 168 kg U_3O_8. Therefore, this loading will be achieved in 168 / 21.1 = 8 hours.

An alternative to RIP is the resin-in-leach (RIL) system where the resin comes in contact with the pulp in the leaching vessel. In certain cases, the ore contains substances such as clays or carbonaceous materials and as the leached metal goes into solution, it is re-adsorbed by these materials thus depleting the solution from the solubilized metal. These materials are called "preg robbers". In order to compete with the preg-robbing and to enhance leaching of the metal, ion exchange resins can be used in the leaching solution (RIL) with the objective to fix the metal before it is absorbed by the preg-robbers. This is encountered in some gold ores. This preg-robbing phenomenon has been experienced in some Lateritic ores in Australia and it has been associated with a high level of clay (Secomb *et al*, 2000). Uranium extraction was increased by 6-14% by the addition of a SBA resin at the start of leaching.

11. Resin fouling

When some ions or some molecules are absorbed by the resin during the absorption cycle but they are not eluted, or only partially, during the elution step, then we say that the resin becomes fouled. When this resin fouling is irreversible then we say that the resin becomes poisoned.

When the resin during the loading period fixes some ions so strongly that the eluent ions cannot displace during elution, or displaces them very inefficiently, then these ions accumulate cycle after cycle to the point where they occupy a significant part of the ionic sites of the resin. As a consequence, the operating capacity of the resin is decreased. In that case, a special treatment should be found that displaces these fouling ions from the resin. If this treatment is too long or too expensive to be applied, then the resin becomes poisoned and it should be replaced when the capacity decrease becomes prohibitive for the process. The time it takes for this to happen depends on the fouling ion concentration in the pregnant solution. As an example, a resin operating at 5 BV/h treats every 24 hours 120 BV. If the fouling ion is found in the feed solution at 1 mg/L level, then every 24 hours they are fixed 120 mg of these ions per liter resin. If the equivalent weight of the fouling ion is for example 60, then the resin fixes 2 meq/L_R every 24 hours. Assuming the resin is in service 50% of the time and that during the elution step all the fixed fouling ions remain in the resin and are not eluted, then in one year the resin will contain some 0.37 eq/L_R of fouling ions, which is about 25% of the total exchange sites of the resin. This corresponds roughly to 5% of the dry weight of the fouled resin. It was assumed here 100% removal of the fouling ions from the pregnant solution and 0% elution during the elution step.

When some molecules are trapped inside the resin or if some precipitation takes place inside the resin and these molecules do not come off during elution then the resin becomes sluggish because the diffusion of the exchanging ions into and out of the resin slows down. In that case, the ionic sites are not occupied so the saturation capacity may not be affected much. However, the leakage during the absorption cycle may increase considerably and the elution may become only partial under normal elution conditions.

Silica

The most frequently encountered resin fouling in uranium recovery is the silica fouling. Scanning electron micrographs (SEM) of resin beads fouled with silica have shown that silica accumulates inside the resin beads in a layer near the surface of the resin (Zaganiaris, 1981).

Silica is found in practically all acid leach liquors at various concentrations, from several ppm up to several grams per liter. The solubility of SiO_2 at the pH values of the acid leach liquors is about 100 ppm. Silicic acid, H_2SiO_3, is a very weak acid and at the low pH values of the pregnant solution it is not ionized. At higher concentrations SiO_2 is stable and remains in solution either as silicic acid or polymerized to dimers, trimers and so on, or as colloidal SiO_2 (Iler, 1979). When the pregnant solution comes in contact with the resin during the absorption period, SiO_2 being non ionic is not excluded by the Donnan potential and penetrates into the resin easily. Once inside the resin, it is found in a high salt concentration and at higher pH than in the outside solution, which are conditions where the rate of gel formation is high (Iler, 1979). Thus, silica polymerizes and forms a gel as soon as it enters the resin, and forms a layer near the surface of the beads. This layer of polymerized silica acts in fact as a barrier to the incoming uranyl sulfate and the outgoing ions so that the kinetics of uranium pick-up slow down with the result of an increased uranium leakage and decreased operating capacity of the resin (fig. 10.1). A similar effect is found in the elution of fouled resins where it is observed that uranium is eluted at a very slow rate and in reasonable residence times the residual uranium on the stripped resin remains high resulting into high uranium leakage and low operating capacity during the following absorption cycle. For example, after 5 BV of H_2SO_4 eluent on a resin fouled with silica at 8% w/w level, only about 50% of the uranium fixed is eluted while on the same resin in new condition, uranium is eluted under similar conditions to better than 95%. Fortunately, silica fouling is reversible and an efficient caustic treatment, as described below, gives to the resin almost its original performance (see figure 10.1 below and the corresponding text, page 223).

The amount of silica loaded on gel type resins depends mainly upon the crosslinking density of the resin. The higher the crosslinking density, the lower the SiO_2 loading on the resin.
When SiO_2 built up in gel type resins reaches a level of about 6% w/w, the operating capacity decrease and the high uranium

leakage become prohibitive and necessitate a cleaning treatment. The most common treatment is with caustic solution as described some paragraphs below.

Macroporous resins also foul with SiO_2 but with a different mechanism (Yahorava *et al*, 2009). The macropores of these resins are filled during the absorption cycle with pregnant solution. SiO_2 therefore penetrates inside the resin beads into the macropores. However, this filling of the macropores does not prevent uranium from being absorbed by the resin (Cable and Zaganiaris, 2010; Rohm and Haas, 2006). It was found that most of the SiO_2 filling the macropores comes off during the H_2SO_4 elution. Subsequently, SiO_2 from the macropores enters the gel phase of the MR resins. It appears however that the overall effect of SiO_2 on resin capacity and leakage is less for MR resins than for gel resins. MR resins are used for a number of years in the former Soviet Union countries in acidic solutions containing moderate SiO_2 levels without any caustic clean up.

Polythionates

The chemical formula of polythionates is $S_xO_6^{2-}$ where x=3,4,5 and 6. They are absorbed on the resin during the absorption cycle and tend to foul the resins. Polythionates fouling can take place in those mines where there is cyanide leaching for gold recovery prior to the acid leaching for uranium recovery (as frequently is the case in South Africa). Polythionates can in this case be formed upon acidification of the gold leaching residues. Polythionates fouling can be minimized by keeping high free acidity in the leach liquor (>5 g H_2SO_4/L) because at this acidity polythionates are less stable.

Cobalt

Again, this kind of fouling can be seen when there is a gold recovery plant before the uranium plant. Fouling is due to the loading of the resin of $Co(CN)_6^{4-}$ which then is not eluted by the common eluents used in uranium elution. One way to avoid this fouling is to wash thoroughly the gold slimes residues to eliminate soluble cyanides and to reduce the amount of cobalt cyanide. Otherwise, most of the cleaning treatments to remove cobalt cyanide from the resin are too costly and too long to be practical. When cobalt accumulates to a level of about 2-2.5% Co, the resin should be replaced.

Molybdenum

As discussed earlier, molybdenum occurs in most uranium minerals and can end-up in the leach liquors in both, acid and alkaline leaching. Molybdenum in the VI state can form complexes of the $MoO_2(SO_4)_n^{2-2n}$ type, as uranium. These complexes however, are more strongly fixed by SBA resins than uranyl complexes and consequently, they tend to accumulate on the resin, thus causing a reduction in operating capacity of the resin for uranium.

Once on the resin, molybdenum can be removed by 5-10% NaOH and therefore it can be removed during the caustic treatment used to remove silica, as described below.

Caustic regeneration

A caustic treatment is applied to clean the resin from silica but also from other foulants such as polythionates, sulfur and molybdenum. For silica removal, the frequency of the treatment varies according to the silica content of the leach liquors, the rate of silica pick up by the resin and the design of the plant and can be from once in a few weeks to once in a year or so. Usually, caustic regeneration is done in a separate column on a spare batch

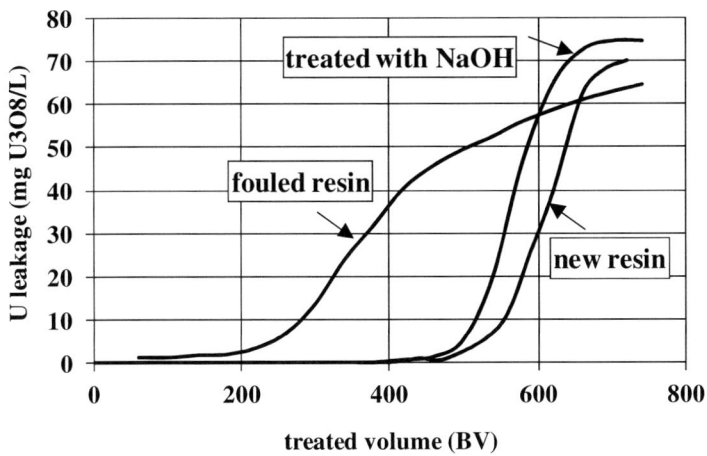

Figure 10.1 Effect of silica fouling on breakthrough curves.

of resin so that the normal operation of the plant is not disrupted. The efficiency of the NaOH treatment in cleaning the resin from silica fouling is shown in figure 10.1. A gel type SBA resin fouled with silica at a level of 8% w/w shows a significant deterioration of kinetics so that a very broad breakthrough curve is obtained. After a NaOH treatment, the kinetics have been improved and the breakthrough curve becomes close to that of a new resin.

In practice, the resin is transferred from the elution column to the regeneration column where it is washed to remove any eluent left. Then it is neutralized with a dilute NaOH solution (about 1%) until the effluent becomes neutral. Then 3 BV of a more concentrated NaOH solution is introduced (about 4-6% NaOH) at a slow flow rate so that the contact time is at least 4 hours, preferably more. Warm temperatures make the treatment more efficient. If time allows, a split regeneration can be used with 3 BV of recycled caustic followed by 3 BV of fresh caustic. The next step is to convert the resin to the SO_4^{2-}/HSO_4^- form before sending it to the absorption column. This can be accomplished using a dilute H_2SO_4 solution (about 7 BV of 1% H_2SO_4) or barren solution provided that the iron level in the barrens is not very high and there is no risk of iron hydroxide precipitation in the resin. In fact, barren solution can be used also as a fluid for transferring the resin back to the absorption column. Higher concentration H_2SO_4 should be avoided not to damage the resin due to excessive neutralization shock and rapid volume change due to resin shrinking.

12. Laboratory resin evaluation

Resin evaluation can be performed by independed institutes, laboratories, equipment manufacturers or process developing companies, frequently for new projects and involving pilot plant studies and including hydrometallurgical aspects. Properties of the ion exchangers such as total exchange capacity, moisture content, physical aspect, commonly called "routine properties", are usually determined by the resin manufacturers or by outside independent laboratories. At a mine, the resin is usually analyzed for uranium and other elements loaded on the resin after the loading and the elution step.

In a uranium mine it is useful to be able to evaluate a resin in a small scale, in addition to the routine resin analysis. These tests can be limited to column tests (loading and elution), hydraulic properties (fluidization) and/or batch tests (equilibrium isotherms, rate of loading, rate of elution). This allows to compare different resins in order to choose the best suited for this particular mine and to follow the evolution of the properties and the performance of the resin used in production in order to better define the operating conditions, apply a clean-up treatment or decide about replacing the resin. A simple set up for column and batch tests would necessitate the following material :

- Peristaltic pump(s) capable to provide flow rates between 0.03 and 2 liters/hour (possibly two different pumps, one for the exhaustion and one for the elution).
- Rubber tubings of different diameters to give the above flow rates
- At least two columns made from glass tubes of an internal diameter of about 2 cm and a height of about 1 meter, by fusing into the tube a sintered glass filter and drawing down the tube at the bottom as shown in figure 11.1
- Vessels and reservoirs to store feed solutions, eluents, eluates and barren solutions
- Mechanical stirrer for batch tests.
- A magnifying glass to observe the physical aspect of the resin sample

In what follows, some guidelines are given how to carry out this kind of tests on resin samples taken from a uranium IER plant.

Preliminary resin preparation

In order that the evaluation of used resins is representative, the sample collected from the column should be representative for the properties to be determined. For column and for batch tests the resin sample should be representative of the whole resin bed. In some continuous systems valves have been installed at the different stages of the loading columns or at the elution columns so that samples of resins or solutions can be taken out. In fixed bed columns, uniform samples can be taken from a manhole at the top of the column using a hollow tube which is pushed through the resin bed (under slight fluidization of the resin bed to facilitate this task) and then plugging the upper part of the tube before taking it out of the resin bed. The resin trapped inside the tube is then emptied into a beaker or another vessel for further tests.

The physical appearance is a relatively easy test which can give very useful information and can solve serious problems. It consists in observing a small resin sample under a magnifying glass or a "microscope" (where the resin sample is in the air). It is examined in order to determine if the resin is clean or if it contains foreign matter like precipitates or suspended matter originating from the feed solution which either cover and stick on the resin beads or are included in the sample between resin beads. The presence of such matter for fixed bed systems may indicate a problem in the filtration before the IX columns, any precipitation inside the IX column or a backwash step that is not performed properly. A backwash in a laboratory column can in that case provide information on backwash conditions that can clean the resin or if a wash is needed, acidic or basic (after ensuring that the column is resistant to such solutions), that can dissolve the foreign matter and clean the resin. Another observation to be done is the broken and cracked beads found in the sample. This is normally expressed as percent of the total number of beads after estimating the number of whole beads that can be reconstructed from the fragments. It is obvious that for this evaluation and for the case of fixed beds, the resin sample should be representative because the resin fragments are usually found accumulated near the top of the bed as a result of backwash operations. If the sample is representative and the broken and cracked beads represent a high fraction of the total, the reasons for this resin breakage should be looked for, for example if it is due to the resin transfer or due to the reagents concentration during the caustic treatment for silica fouling.

For column or batch tests, the resin sample should be taken after the elution column and before the resin goes to the absorption column. In this way, the resin is ready for evaluation and no other conditioning is needed. Fresh SBA resins are usually produced and supplied in the Cl⁻ form except special products such as SBA resins in SO_4^{2-} or CO_3^{2-} form for acid or carbonate leach uranium mines.

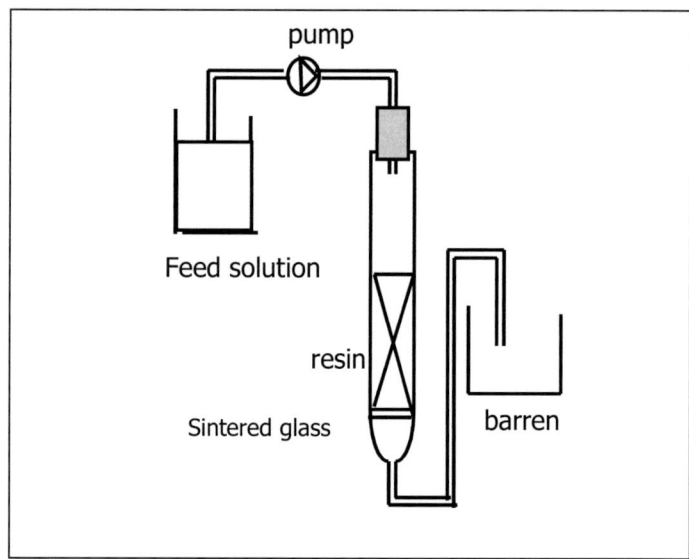

Figure 11.1 Column set up for operating capacity determination

Consequently, if fresh resin is included in the test for comparison, it has first to be converted to the same ionic form as the used resin sample. This can be done by placing the fresh resin into a column and passing an excess of eluent solution through the resin. A column set up is illustrated in figure 11.1. If two columns in parallel are needed, then the pump should be equipped with two heads, one for each column.

Column tests

Charging Ion Exchange Resins Into Columns

Deionized water should be used to transfer ion exchange resins to the column. If soften or tap water is used, the resin will be converted partially to the ionic form reflecting the salts dissolved in the water. If the salt concentration is not very high, the use of softened or tap water can be acceptable. The column itself should contain some water at the start of the operation, and the ion exchange resin should be poured into the column as water slurry, using a beaker or other vessel.

Sufficient resin should be placed into the column so that the resin bed height is about 50 cm. With a 2 cm ID column, this means about 150 ml of resin. Ion exchange resins should always be fully hydrated before introduction into the column. The following steps, then, should be followed in charging resins properly:

1. Add resin-water slurry as described above to a column containing some water.

2. Occasionally drain excess water through the bottom of the column.
3. Do not permit the liquid level to fall below the resin level.
4. Continue adding in this manner until all resin is transferred.
5. Do not load columns to more than about half their height.
6. Introduce deionized water upflow very slowly. Increase the flow until the bed of resin expands to near the top end of the column. Maintain this flow until all air pockets are removed and all the particles have achieved mobility. (Extremely small particles may be allowed to exit the column.) This step is most important and, if performed correctly, will result in good particle size classification with the smaller particles at the top and the larger ones at the bottom portion of the column.
7. Stop the flow of water and allow the resin to settle by gravity.
8. Pass water downflow at a linear velocity of about 20 m/h. After a few minutes and making sure the resin height remains stable, measure this resin height and calculate the resin volume using the column cross-section.

Exhaustion (loading) cycle

Before proceeding to the exhaustion cycle, consideration should be given to the flow rates to be employed, concentrations of the solutions that will contact the resin and the volume of these solutions required. If pregnant solution from production is to be used, a full analysis should be available. If the plant is equipped with fixed bed columns, the specific flow rate (BV/h) used in the plant should be used in the laboratory tests. If two columns in series are involved, a Bed Volume is taken as the sum of the volumes in the two columns. Otherwise, a typical value of the specific flow rate to use is 5 BV/h. With 150 ml of resin in the column, this means 750 ml/h flow rate. With a 2 cm ID column, this results in a linear velocity of 2.4 m/h. This linear velocity will most probably be different from the linear velocity in the plant, however for the case of uranium recovery where the kinetics are particle diffusion controlled, it is the specific flow rate that should be matched with the plant flow rate in the laboratory tests.

With a column test one can achieve the following objectives: determine the saturation capacity, measure the uranium leakage and estimate the ion exchange zone length that corresponds to the operating conditions.

Since the resin samples come from the elution column after elution, then the obtained uranium leakage would be similar to the leakage that the plant should experience.

The IEZ helps in confirming the design of a two-columns-in-series configuration. If the IEZ is less than the resin bed height of one column this assures that at the breakthrough, the first column will be saturated. It also can be useful in moving packed bed systems where the IEZ should be smaller than the resin height less the height of the resin slag that is transferred to elution. Since the IEZ length depends on the linear velocity of the solution and in the laboratory column tests it is not the same as in the plant, the IEZ should be calculated as the fraction of the total resin bed height.

The estimate of the IEZ is illustrated with the following example (figure 11.2)

Assume that with a pregnant solution containing 100 mg U_3O_8/L a column test was run and the breakthrough curve of figure 11.2 was obtained. At the end of the cycle, after more than 600 BV of solution had been treated, the resin saturation capacity was found to be 45 g U_3O_8/L_R. This saturation capacity can be calculated from the total amount of uranium passed through the resin less the amount of uranium leaked through until the end of the cycle. This capacity of 45 g U_3O_8/L_R represents 450 BV of solution (=45 g U_3O_8/L_R*1000/100 mg U_3O_8/L). The IEZ is represented in the figure by $\Delta(BV)$ and here it is 200 BV. The resin bed fraction that occupied the IEZ is then 200 / 450 or 44%.

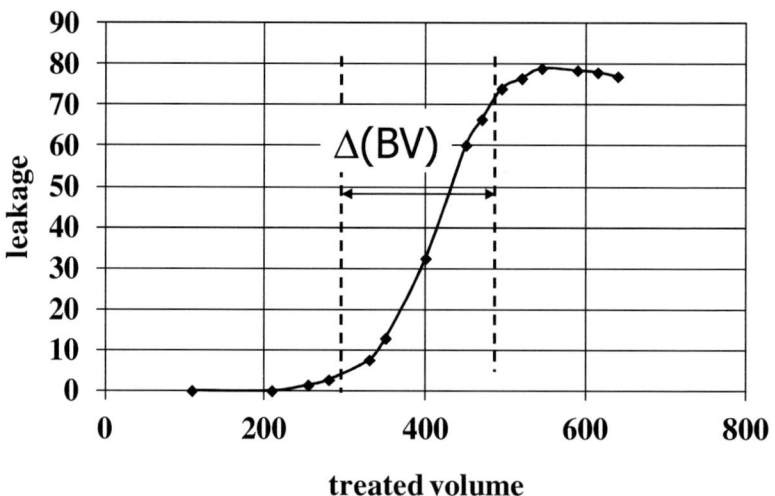

Figure 11.2 Determination of the Ion Exchange Zone (see text)

If this trial intends to simulate a two-columns-in-series arrangement then the $\Delta(VB)$ becomes 400 BV since the volume of the resin in each column is one half of the total. The fraction of the resin height in the first column that occupies the IEZ is then 400 / 450 or approximately 90 %. At the breakthrough from the second column, the first column would be over-saturated by about 10%.

The breakthrough of the run illustrated in fig. 11.2 takes place after about 260 BV. In this cycle however, 100% of the resin at the beginning of the loading cycle was found in the ionic form that corresponds to the eluent. In a two-columns-in-series system on the other hand, in the second cycle the first column would go to elution and the second column, being partially exhausted, would become first column. Since the column that goes to elution in our example is exhausted by 100%, or loaded with 45 g U_3O_8/L_R, the loading cycle would last 450 BV (for one column) or 225 BV for the total resin in both columns. If we represent therefore the cycle of two columns in series, in a figure like the 11.2, the breakthrough would start at 225 BV while the IEZ would remain the same. At the specific flow rate of 5 BV/h, the cycle time is expected to be about 45 hours (=225/5).

As it is seen, with one column test as described here one can evaluate a resin for a two-columns-in-series merry-go-round system. Similar tests can be run to evaluate a resin for a moving packed bed system. In that case, a column run at the same specific flow rate as in the plant should give an IEZ of 80% or less of the resin bed in order to have saturation of the resin layer that is transferred to elution.

If the mine employs a fluidized bed continuous IX system or a RIP system, a fixed bed column trial cannot directly predict the performance of the resin in those systems. In these cases batch tests to determine equilibrium isotherms and rates of loading would help, using the appropriate model to calculate the number of stages and compare to those in the plant. Instead, a fixed bed column test can be used for comparing the used resin with a new one.

Elution

When the elution system in the plant is continuous, that is the resin is transferred to the EC in slags and the eluent flows counter-current to the resin, we have the following situation :
Assume that the elution system consists of 5 columns in series (or one column with 5 stages). One slug of 4 m3 of resin is transferred every 3 hours. The whole elution circuit would then contain 5*4 = 20 m3 resin and each slug of resin remains in the circuit for 3*5 = 15 hours. The resin flow rate is 4/3 = 1.33 m3/h. Assume now that the eluent flows at an average flow rate of 5 m3/h. The number of Bed Volumes of eluent that have passed through the elution circuit from the time one resin slug goes to elution to the time this slug goes to absorption is 5*15/20 = 3.75. The specific flow rate of the eluent is 3.75/15 = 0.25 BV/h.
A continuous elution system like the one described here can be approximated by the elution of a fixed bed column using the same number of BV of eluent and the same specific flow rate. In the example of the resin used above, with 150 ml of resin in the column, the flow rate of the eluent would be 37.5 ml/h. It will take 15 hours to pass 3.75 BV. Following the 3.75 BV of eluent, the resin is rinsed with water and all effluents are collected for analysis. The residual uranium on the stripped resin can be determined by passing additional eluent to a large excess and analyze for uranium.

Batch experiments

Batch experiments include equilibrium isotherms and kinetic data for loading and elution. The equilibrium and the rate of loading data are useful for CIX systems such as the NIMCIX system and the RIP. These systems consist of a number of contactors or stages with the pregnant liquor passing through these contactors, from the first to the last one, until the first one reaches saturation. The time to saturation of the first contactor is called cycle time. At this time, the resin in the second contactor would have reached equilibrium with the liquor flowing from the first contactor. Similarly, the resin in the subsequent contactors comes in equilibrium with the liquor coming out from the previous contactor. The calculations of these equilibria was discussed in Chapter 10, page 198. Data obtained from the batch equilibrium experiments provide equilibrium values whilst the data obtained from kinetic experiments enables one to calculate to what degree the equilibrium is reached during a certain cycle time period.

Equilibrium tests

In the construction of equilibrium isotherms, like those of figure 4.2, page 82, two ions are involved and the isotherm consists in plotting the resin loading for one of them versus the concentration of that ion in solution. Constructing equilibrium curves for uranium, either representing acidic or basic leach conditions, there may coexist more than two ions, for example in acid leach we have SO_4^{2-}, HSO_4^-, and the various uranyl sulfate complexes, plus eventually $Fe(SO_4)_2^-$ etc. for alkaline leach we have at least CO_3^{2-}, HCO_3^- and $[UO_2(CO_3)_3]^{4-}$. For that reason, by batch equilibrium tests here it is meant the construction of curves of uranium loading on the resin versus uranium concentration in solution under certain solution composition that represents the pregnant solution composition in the plant. For example, the curve in figure 7.3, page 160, gives uranium loadings in equilibrium with a solution of a certain initial composition and pH. The curve has been constructed by bringing together a given volume of solution and a given volume of resin and allowing enough time (usually overnight) under agitation, for example in a shaker, to reach equilibrium. After that, the solution is analyzed for uranium content. The solution and resin volumes are chosen so that the whole range of concentrations is covered. Another way to construct equilibrium curves is to prepare solutions having different uranium concentrations by starting from the same initial solution and dissolving different quantities of uranium and allow to pass in a column through a given quantity of resin a large excess until the resin is saturated, that is, until the influent and the effluent compositions are the same. Then take the resin out and determine the uranium loading by elution. This, for each solution made with different uranium concentrations. This approach is more precise concerning the composition of the solution at equilibrium, it is however more time consuming.

Rate of loading tests

In this test a small quantity of resin is placed in contact with a relatively large quantity of solution and periodically a solution sample is taken out and analyzed for uranium. A typical experiment consists in placing a 10 ml resin sample into a 2 liter solution under agitation using a mechanical stirrer and take a 25 ml sample at given time intervals. Magnetic stirrers are to be avoided because they grind the resin and therefore the particle size of the resin sample, and the kinetics behavior, will change during the test. After the uranium analysis in the samples, the solution volume and the remaining uranium in the solution is adjusted to reflect the removal of the 25 ml sample. Then, the uranium loading on the resin sample is calculated. In order to interpret the results, one can simulate the uranium take-up by the resin by using a chemical kinetics model. In this case, we assume a first order kinetics for the resin loading :

$$dy/dt = k*x*y \quad \text{or} \quad dy/y = k*x*dt \quad (11.1)$$

where y is the uranium concentration on the resin, x is the uranium concentration in solution and k a rate constant. By integrating eq. 11.1 from y_o to y_{eq} and from time t_1 to t_2, the time interval and taking x the average value of x during the time interval, x_{ave}, we have :

$$\ln \frac{(y_{eq} - y_o)}{(y_{eq} - y)} = k\, x_{ave}(t_2 - t_1)$$

Or

$$y = \frac{y_{eq}(\exp(kxT) - 1) + y_o}{\exp(kxT)}$$

where x is x_{ave} and $T = t_2 - t_1$

we then have:

$$k = \frac{1}{x_{ave}(t_2 - t_1)} * \ln\frac{(y_{eq} - y_o)}{(y_{eq} - y)} \qquad (11.2)$$

It is this rate constant that can be used in the calculation of the number of stages in a continuous fluidized bed system discussed in Chapter 10, page 199. The time T is in that case the cycle time, that is the time between resin transfers.

It may come from the results of the rate of loading tests that the rate constant is not really constant but that it depends on the resin loading: as the resin loading increases, the rate constant may decrease.

Figure 11.3 illustrates a rate of loading test for a gel-type SBA resin in contact with a solution containing initially 22 g SO_4/L, 2.5 g Fe^{3+}/L, 180 mg U_3O_8/L at a pH of 1.7. The test conditions were as described above.

The calculation of the rate constant k is done in this kind of test from the values of the concentration of uranium in the solution at time t, x(t), from which it is calculated the average value of x between two consecutive values of time, x_{ave}, , then it is calculated the total amount of uranium fixed on the resin during this time interval, y(t), as well as the equilibrium value of uranium on the resin that corresponds to the x_{ave}, $y_{eq}(t)$. The rate constant k is then calculated using equation 11.2.

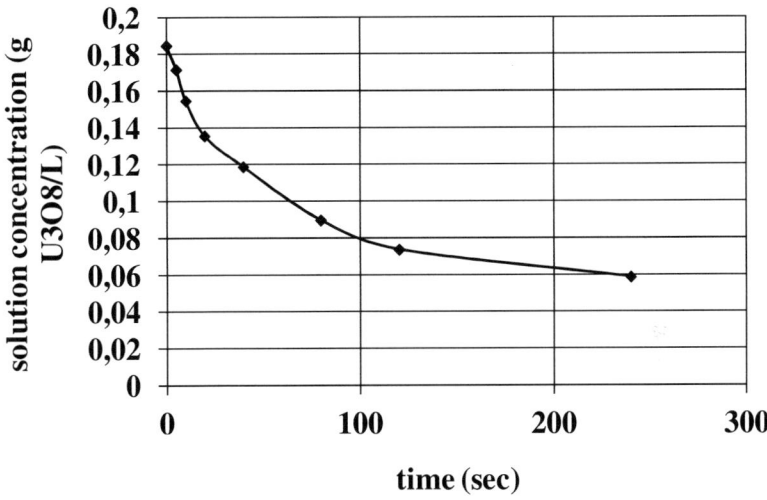

Figure 11.3 Kinetic test. Gel-type SBA resin in contact with a solution containing initially 22 g SO_4/L, 2.5 g Fe^{3+}/L, 180 mg U_3O_8/L at a pH of 1.7

REFERENCES

Ahrland S (1951). On the complex chemistry of the uranyl ion. 5. complexity of uranyl sulfate. Acta Chem Scand, **5** : 1151-1167

Alexandratos SD, Brown GM, Bonnesen PV, Moyer BA (2000). Bifunctional anion-exchange resins with improves selectivity and exchange kinetics. *US Patent 6,059,975.*

Applebaum SB (1968). *Demineralization by Ion Exchange*. Academic Press, New York

Arden TV (1968). *Water Purification by Ion Exchange.* Butterworth, London.

Arden TV, Wood GA (1956). Adsorption of Complex Anions from Uranyl Sulfate Solutions by Anion-exchange Resins. *J Chem Soc,* **1956:** 1596-1603.

Arden TV, Nolan JD, Gower BJ, Wright FA (1959). Recovery of thorium from sulphate solutions by anion exchange. *J Appl Chem* **9:** 406-409.

Ballestrin S, Low R, Reynaud G, Crane P (2014). Honeymoon mine Australia: Commissioning and operation of the process plant using a novel solvent extraction reagent mixture in a high chloride environment. *ALTA 2014 Uranium-REE Conference, Perth, Australia, May 24-31, 2014*

Bester J, Corbet S, Delameilleure S, Zaganiaris E (2016). Uranium recovery from H2SO4 eluates. *SCI Conference IEX2016, Cambridge UK, July 6-8 2016.*

Boari G (1974). Method for removing sulfate and bicarbonate ions from sea water or brackish water through the use of weak anionic exchange resins containing amino groups of the primary and secondary type. *US Patent 3,842,002 A.*

Bortnick NM (1962). Catalyzing reactions with cation exchange resin. *US Patent 3,037,052.*

Boytsova T, Babain V, Korolev V, Ozawa M, Suzuki T, Pokhitonov Y (2011). Combined separation of Pd and Tc from the raffinates of spent nuclear fuel reprocessing. *7th International Symposium on Technitium and Rhenium.* Moscow, 2011, p. 59

Britt PF, Gill GA, Schneider E (2014). Overview of Fuel Resources Program: Seawater Uranium Recovery Sponcored by the US Department of Energy. *IAEA Symposium on Uranium Raw Material for the Nuclear Fuel Cycle, Vienna, June 27, 2014.*

Brunauer S, Emmett PH and Teller E (1938). Adsorption of gases in multimolecular layers. J Am Chem Soc, **60** : 309-319

Cable PI and Zaganiaris E 2010. Process for uranium recovery using anion exchange resins. *US Patent 7,655,199 B2*

Carmen C, Kunin R (1986). Recovery of uranium from carbonate leach liquors using weak-acid cation exchange resins. *Reactive Polymers* **4:** 77-89.

Carr J, Zontov N and Yamin S (2008). Meeting the future challenges of the uranium industry. *ALTA 2008 Uranium Conference, June 19-20 2008, Perth, Australia*

Carr J, Chamberlain T, Zontov N (2012). Method and system for extraction of uranium using an ion-exchange resin. *Patent application WO 2012109705 A1.*

Chanda, M and Rempel, GL, (1992). Uranium sorption behavior of a macroporous, quaternized poly(4-vinylpyridine) resin in sulfuric acid medium. *Reactive Polymers*, **18**, 141-154.

Clifford DA and Weber Jr WJ (1983). The determinants of divalent/monovalent selectivity in anion exchangers. *Reactive Polymers* **1:** 77-89.

Clifford DA (1999). Ion Exchange and Inorganic Adsorption. *In* Letterman R *Water Quality and Treatment*. McGraw Hill, Inc., New York, 5th Edition, 9.1-9.91.

Cloete FLD, Haines AK (1971). S.A.Patent Appl. 71/3139

Cloete FLD, Streat M (1967). Brit. Patent 1 070 251

Corte H, Meyer A (1972). Anion exchanger with sponge structure. US Patent 3,637,535.

Dale JA, Irving J (1992). Comparison of strong base resins types. *In* Slater MJ, *Ion Exchange Advances: Proceedings of IEX'92*, 33-40.

DAlelio GF (1952). Ion exchange resins from a vinyl pyridine or a vinyl quinoline and a vinyl ethinyl hydrocarbon. US Patent 2,623,013 A.

Davankov VA and Tsyurupa MP (1989), Structure and properties of porous hypercrosslinked polystyrene sorbents 'Styrosorb', Pure and Appl. Chem. **61**: 1881-1888

de Dardel F and Arden TV (1989). Ion Exchangers. *In Encyclopaedia of Technical Chemistry.* VCH, Weinheim, Germany Vol. A14

Dow Technical Information "Using Ion Exchange Resins Selectivity Coefficients" *Form N° 177-01755-0207*

Dow Technical Bulletin (2016). Addressing Pressing Needs for Wastewater Treatment and Contaminant Removal. *Form N° 177- 03571, May 2016*

Drozdzak J, Leermakers M, Gao Y, Phrommavanh V, Descostes M (2016). Novel speciation method based on diffusive gradients in thin films for in situ measurement of uranium in the vicinity of the former uranium mining sites. *Environ Pollut* **214:** 114-23.

Dunn G and Teo YY (2015). Recovery of uranium from a strong sulfuric acid loaded strip or eluate. *ALTA 2015 Conference, Perth, Australia, May 2015.*

Ford M (2009). The evolution of uranium extraction technology. *MINTEK's 75th Anniversary Conference, June 4, 2009.*

Girsh DJ, Martinez H, Miller M (2001). Methylene bridged resins yield enhanced osmotic stability in contacting with organic acid streams. *J Separation Science* **24** *: 477-478*

Greer AH, Mindler AB, Termini JP (1958). New ion exchange resin for uranium recovery. Ind Eng Chem 50(2): 166-170.

Gu B, Brown GM, Alexandratos SD, Ober R, Patel V (1999). Selective anion exchange resins for the removal of perchlorate ClO4- from ground water. *ORNL-TM 13753 report.*

Gupta CK, Singh H (2003). *Uranium Resource Processing, Secondary resources.* Springer-Verlag Berlin. P. 156-168.

Haines AK (1972). *S.A.Patent Appl. 72/4328.*

Haines AK, Craig WM, Faur A, Hendriksz AR, Wills WJ and Nicol DI (1975). The use of weak base resins in the recovery of uranium from unclarified leach liquors. *11th International Mineral Processing Congres*, Rome, paper 32.

Haines AK (1977).The South African Progrmme on the Development of Continuous Fluidized Bed Ion Exchange with specific Reference to its Application to the Recovery of Uranium. Presented at the *Process Development for the treatment of ores and residues for recovery of uranium and other values*, Orange Free State Colloquium, 16th and 17th November 1977

Hanson SW, Wilson DB, Gunaji NN (1987). Removal of uranium from drinking water by ion exchange and chemical clarification. *Water Engineering Research Laboratory, Cincinnatti, Ohio.*

Hassid M, Ketzinel Z, Volkman Y (1986). Recovery of uranium from wet-process phosphoric acid by liquid-solid ion exchange. *US Patent 4,599,221.*

Helfferich F (1962). *Ion Exchange.* McGraw-Hill Book Company, Inc.

Helfferich F (1966) Ion-exchange kinetics. *In* Marinski J *Ion exchange, a series of advances*, 65-100

Hennig C, Schmeide K, Brendler V, Moll H, Tsushima S, Scheinost AC (2007). The structure of uranyl sulfate in aqueous solution-monodentate versus bidentate coordination. *AIP Conference Preceedings* **882**, 262-264

Hubicki Z, Kołodyńska D (2012). Selective removal of heavy metal ions from waters and waste waters using ion exchange methods. *In* Kilislioglu A *Ion Exchange Technologies*. InTech, 193-240.

IAEA (2001). Manual of acid in situ leach uranium mining technology. *IAEA-TECDOC-1239, Vienne Austria*

Iler RK (1979). *The Chemistry of Silica*. John Wiley & Sons

Jamrack WD (1963). Rare metal extraction by chemical engineering techniques. Pergamon Press p. 101-110.

Kabay N, Demircioglu M, Yayli S, Günay E, Yüksel M, Saglam M, Streat M (1998). Recovery of uranium from phosphoric acid solutions using chelating ion-exchange resins. *Ind Eng Chem Res* **37**: 1983-1990.

Kudlanov VN (2000). Perfection of uranium desorption process stage and processing of natural uranium chemical concentrate technical report (in Russian). Presented at the Conference *Actual problems of uranium deposits mining by in-situ leaching method* , Almaty, Kazakhstan, 2000.

Kunin R (1958). *Ion Exchange Resins*. John Wiley, 2nd Edition.

Kunin R, La Terra T (1986). Uranium recovery from carbonate leach liquors using carboxylic acid cation exchange resin. *US Patent 4,606,894.*

Ladeira ACQ, Gonçalves CR (2007). Influence of anionic species on uranium separation from acid mine water using strong base resins. *J Hazard Mater* **148:** 499-504

Ladeira ACQ, Sicupira LC (2013). Application of ion exchange resins to recover uranium from acid mine drainage. *Proceedings of the 13th International Conference of Environmental Science and Technology, Athens, Greece, 5-7 September, 2013.*

Masuda S, Andou K, Kubota H, Watanabe J (2000). Thermally stable anion exchange resin (Diaion® XSA series): characteristics and applications. *In* Greig JA, Editor. *Ion Exchange at the Millennium.* SCI, Imperial College Press, 253-260.

McGarvey FX and Ungar J (1981). The influence of resin functional group on the ion exchange resin recovery of uranium. *J South African Institute of Mining and Metallurgy* April 1981, 93-100

McGarvey FX and Hauser EW (1982). Kinetic studies on gel and macroporous anion exchangers using the uranyl sulfate/sulfate exchange. Presented at : *NATO Advanced Study Institute, Mass Transfer and Kinetics of Ion Exchange*, Maratea (Pz), Italy, May 31-June 11, 1982

McIntosh AM, Viljoen EB, Craig WM, Taylor JL (1982). The design, commissioning and performance of the NIMCIX section of the Chemwes uranium plant. *J S Afr Inst Min Metall* **82 (7):** 177-185.

Mikhaylenko M, van Deventer J (2009). Notes of practical application of ion exchange resins in uranium extractive metallurgy. *ALTA 2009 Uranium Conference, Perth, Australia*

Naidoo A., Archer S.J. and Coetzee V.E. The recovery of acid from a uranium ion exchange eluate using nano-filtration technology. Africa Australia Technical Mining Conference, June 11-12, 2015, Adelaide, Australia.

National Institute for Metallurgy (1978). *the NIMCIX contactor* Continuous ion exchange in hydrometallurgy, technical bulletin N° 2

Ogden MD, Soldenhoff K (2013). The role of chelating resins in uranium processing. *ALTA 2013 Uranium-REE Conference, Perth, Australia, May 25-June 1st, 2013*

Patrin AP (2005). Perfection of uranium adsorption and ion exchange resin regeneration process stages. *In* M.I.Fazlullin *"Underground and heap leaching of uranium, gold and other metals"*, "Minerals and Metals" publishing House, Moscow, Vol. 1, p. 277-280 (in Russian)

Peacock M, McDougall S, Boshoff P, Butcher D, Ford M, Donegan S, Bukunkwe D (2016). Paladin Energy Ltd-Nanofiltration technology for reagent recovery. *ALTA 2016 Conference, Perth, Australia, May 21-28, 2016.*

Preuss AF and Kunin R (1958). Uranium recovery by ion exchange. *In* Clegg JW and Foley DD *Uranium ore processing.* USAEC, Addison-Wesley Pub. Co., Ch. 12.

Porter, RR (1971). *S.A.Patent Appl. 71/8632*

Rezkallah A (2012). Method for the recovery of uranium from pregnant liquor solutions. *Patent application US20120125158 A1.*

Rezkallah A, Ferraro JF, Pellny PM (2016). Removal of uranium from wáter. *European Patent EP 2 931 424 B1*

Rezkallah A, Dunn G.M. Method for the recovery of uranium from a strong sulfuric acid loaded strip or eluate. WIPO Patent Application WO2015/135017, 17.09.2015

Riegel M, Höll WW (2008). Removal of natural uranium from groundwater by means of weakly basic anion exchangers. *In* Cox M, *Recent Advances in Ion Exchange Theory and Practice.* (proceedings of IEX2008). SCI, 2008, 331-338.

Rohm and Haas Company (2006). *AMBERLITE^{TM} IRA910U Cl and AMBERSEP^{TM} 920U Cl Macroreticular Strong Base Anion Resins for Uranium Recovery from Rohm and Haas*

Rohm and Haas Company (1980). *Amberlite Ion Exchange Resins for Uranium Hydrometallurgy.*

Rosenberg E, Pinson G, Tsosie R, Tutu H, Cukrowska E (2016). Uranium Remediation by Ion Exchange and Sorption Methods: a Critical Review. *Johnson Matthey Technol Rev* **60(1):** 59-77.

Secomb R, Ring R, Macnaughton S (2000). Enhancing uranium recovery using resin-in-leach technology. *Uranium 2000: proceedings of the Internations Symposium on the Process Metallurgy of Uranium: September 9-15, 2000, Saskatoon, Sask, Canada, 429-453.*

Slater MJ (1991). *The principles of Ion Exchange Technology.* Butterworth Heinemann.

Soderholm L, Skanthekumar S, Wilson RE (2011). Structural correspondence between uranyl chloride complexes in solution and their stability constants. *J Phys Chem A 2011,* **115:** 4959-4967.

Soldenhoff K, Wilkins D, Shamieh M, Ring R (2000). Solvent extraction of uranium in the presence of chloride. Uranium 2000, Proceedings of the International Symposium on the Process Metallurgy of Uranium. C2000 339-351, Canadian Institute of Mining, Metallurgy and Petroleum, Montreal, Quebec, Canada.

Soldenhoff K, Tran MT, Griffith C (2009). Recovery of uranium from phosphoric acid by ion exchange. IAEA Technical Meeting, Uranium from Unconventional Resources, Vienna, 2009.

Tebibel P and Zaganiaris E (1992). Nitrate removal from water : the effect of structure on the performance of anion exchange resins. *In* Slater MJ *Ion Exchange Advances*. Elsevier Applied Science, 41-48

Tsouris C, Liao W-P, Das S, Mayers R, Janke C, Dai S, Kuo L-J, Wood J, GillG, Flicker Byers M, Schneider E (2015). Adsorbent alkali conditioning for uranium adsorption from seawater. ORNL-LTR 2015/568

Udayar T, Kotze MH, Yahorava V (2011). Recovery of Uranium from dense slurries via resin-in-pulp. *Proceedings 6th South African Base Metals Conference 2011, South African Institute of Mining and Metallurgy,* 49-64.

Udayar T, Yahorava V, Kotze MH (2013). Approaches for rejection of impurities in uranium ion exchange. *ALTA 2013 Uranium-REE Conference, Perth, Australia, May 25-June 1st, 2013*

van Deventer J (2013). Ion exchange resins for uranium recovery: The durability question explored. *ALTA 2013 Uranium-REE Conference, Perth, Australia, May 25-June 1st, 2013* van Hege B, van Tonder D, Bell R, Wyethe J and Kotze M (2006). Recovery of base metals using MetRIX^{TM}. *ALTA Conference, May 2006, Perth, Australia*

Volkman Y (1987). Recovery of uranium from phosphoric acid by ion exchange. In The recovery of uranium from phosphoric acid. *IAEA Tecdoc-533, IAEA, Vienna 59-68.*

Wang D, Chen ASC, Lewis GM (2008). Arsenic and uranium removal from drinking water by adsorptive media. US EPA demonstration project at Upper Bodfish in Lake Isabella CA. *National Risk Management Research Laboratory, Cincinnati, Ohio.*

Wright, RS, (1979). The prediction of concentration profiles for a NIMCIX column absorbing uranium from acqueous solution, *NIM publication No. 7131, National Institute for Metallurgy, Randburg, South Africa*

Yahorava V, Scheepers J, Kotze MH, Auerswald D (2009). Impact of silica on hydrometallurgical and mechanical properties of RIP grade resins for uranium recovery. *J South African Institute of Mining and Metallurgy* **109:** 609-619

Zaganiaris EJ (1981). Effect of physical and chemical structure of ion exchange resins on silica fouling in acid leach liquors in uranium hydrometallurgy. *Hydrometallurgy 81*, Proceedings of a SCI Symposium, E4/1, Manchester, 1981.

Zhu Z, Pranolo Y, Cheng CY (2013). Uranium recovery from alkaline leach solution using ionic liquid Cyphos 1L-101. *ALTA 2013 Uranium-REE Conference, Perth, Australia, May 25-June 1st, 2013*

Zhu Z, Tulpatowicz K, Pranolo Y, Cheng CY (2014). Uranium recovery from sulfate leach solution containing chloride using a mixed system of D2EHPA and Cyphos IL-101. *ALTA 2014 Uranium-REE Conference, Perth, Australia, May 24-31, 2014*

Zontov N (2006). Continuous counter current ion exchange in uranium ore processing. *ALTA 2006 Uranium Conference, Perth, Australia*

Subject Index

A

AAC Pumpcell	209, 213, 214, 216, 217
Acid leach	36, 142, 145, 151-155, 163, 178, 181 193, 194, 219, 221
Alkaline leach	140, 142, 164, 178-181, 183, 191, 194, 222
Alamine® 336	142, 163
Amidoxime resins	36, 187-189
Aminomethylphosphonic (AMP) resins	20, 36, 145, 162, 186
Ammine complexes	33
Ammonium diuranate	141, 144
Ammonium sulfate elution	166, 167
Apparent density	45, 46
Apparent volume	43, 44, 45
Attrition (resin)	61, 209, 211, 213
Average pore diameter	26, 59-60

B

Batch operation	101-111, 216
Bed Volume (BV)	108, 114, 231
Breakthrough curves	115-119
Broadening front	113, 118
Bufflex	141, 171

C

Carbonate leaching: see alkaline leaching	
Carnotite	145, 178, 179
Caustic regeneration	220, 221, 223
Chelating resins	35, 37, 83, 145, 171, 172, 186
Chloride elution:	
Acid leach	166
Alkaline leach	181
Chlorides (effect of)	161
Chloromethylation	30
Clean-iX® cRIP system	206, 207, 211, 212, 213
Cobalt fouling	222
Co-ions	73-76
Complex-ion equilibrium	146
Continuous ion exchange	110, 145, 195
Core-shell model	99
Counter-ions	22, 39, 42, 73, 76, 88, 89, 90

D

D2EHPA	37, 163, 186
Density (resin)	45, 198
Apparent density	45, 46
Crosslinking density	39, 41, 62, 76, 170, 220
Polymer censity	39, 59
Pulp density	217
Skeletal density	46
True wet density	43, 46, 59
Diffusion coefficient	95, 98, 99, 156
Diphonix™ resins	37, 145
Dissociation constant	69, 70
Distribution coefficient	150

Donnan potential   74, 75, 76, 146, 154, 220

E

Effective size (ES)   48
Eluex   141, 171
Equilibrium
    Complex-ion   146
    Constant   78, 84, 110
    Favorable   79, 80, 83, 108, 113, 121, 122
    Ion exchange   73-85
    Isotherms   59, 79, 80, 81, 82, 85, 88, 155, 166, 173, 198, 225, 233, 235
    Unfavorable   79, 80, 83, 105, 106, 125
Ergun equation   49, 51, 52, 53
Expansion (resin bed)   43, 44, 55, 56, 57, 128, 192, 198

F

Fick's law   96
Film diffusion controlled kinetics   95, 97
Fixed bed column operation   111, 124, 193, 201, 203, 226
Fluidization   47, 55-58, 124, 192, 198, 200, 225
Fluidized bed   196-201
Fouling   218
Free water   41, 42, 64

H

Harmonic Mean Size   49
Heap leaching   138, 139
Hoffman degradation   65, 66
Hypercrosslinked (adsorbents)   26, 27

I

Iminodiacetic (IDA) resins	20, 35, 83, 162
In-situ-leaching (ISL)	138, 139, 153
Ion exclusion	75, 76, 77
Ion exchange isotherms see equilibrium isotherms	
Ion exchange zone (IEZ)	112, 115, 117, 120, 121, 123, 124, 131, 231-233
Iron	141, 142, 152, 166, 176, 186
Iron (cake)	193
Iron (effect on operating capacity)	163, 164

L

Leakage (simulations)	118, 123

M

Macronet	26
Macroporous	25, 62, 100, 165, 176
Macroreticular see macroporous	
Merry-go-round	131-134, 190, 191, 233
MetRIX^{TM}	213
Moisture	38-43, 46, 51, 52, 60, 99, 156, 166, 169, 170, 180, 211, 225
Molecular Recognition Technology	37
Molybdenum	141, 164, 165, 180, 222, 223
Monodisperse particle size	49

N

Nanofiltration (NF)	171
Nernst equation	75

Nernst film	95, 122
Nernst-Planck equation	97
NIMCIX column	195-197, 199, 235
Nitrates (effect on capacity)	161
Nitrate elution	
Acid leach	141, 166, 201, 205
Alkaline leach	181

O

Operating capacity	121, 122
Oxidation	67-68

P

Particle diffusion controlled kinetics	95, 98, 99, 122, 123, 153, 156, 231
Particle size	47-50
Effect on capacity (acid leach)	156-158
Effect on breakthrough curves	122-123
pH (effect on capacity)	160-161
Physical degradation	61
Polyphenols impregnated resins	179
Polythionates (fouling)	221
Porogen	24-26, 59
Porter design	195, 200
Preg robbing	213
Pressure drop	50
Pyridine resins	34, 155, 159, 161-163

R

Rate constant	199, 237-238
Rate of loading	97, 225, 237-238

Recycling (eluent)	144, 165, 193
Resin-In-Leach	213
Resin-In-Pulp	139, 145, 165, 209-216, 233
AAC Pumpcell	209, 213-217
Clean-iX®	206, 207, 211, 212, 213
MetRIX™	213
Routine properties of resins	225

S

Salt splitting	70, 72
Scrubbing	141, 144, 169, 201, 203
Seawater (uranium from)	187-189
Selectivities	86-94
Selectivity coefficient	77, 78, 88, 91, 101, 104, 105, 116, 117, 120, 121, 148, 161, 169,
Selectivity sequence	35, 36, 86, 89, 92, 140
Self-sharpening front	112, 113, 114, 118, 120
Separation factor	77-83, 86, 89, 93, 98, 101, 112, 116, 118, 119, 125
Silica fouling	109, 154, 195, 196, 219-220, 223-224, 228
Shrinking core model	99
Skeletal density: see density	
Sodium bicarbonate elution	181
Sodium diuranate	141, 144, 145, 179
Solids fraction	39, 46
Solvent extraction (SX)	141-145, 162, 163, 171, 186, 194
Solvent impregnated resins (SIR)	37
Stability constant	146, 173, 174, 178
Stirred tanks model	115
Stoke's law	46, 55
Sulfuric acid elution	166-167, 170-171
Swelling	38-43, 61, 62, 63, 86, 99

T

True Wet Density: see density
Thermal degradation   64-67
Thorium   164
Type 1 SBA resins   30, 65, 66, 91, 92, 155, 160, 170, 176
Type 2 SBA resins   30, 66, 92, 155, 176

U

Uniformity coefficient (UC)   48
Uraninite   137, 178
Uranium peroxide   141
Uranyl sulfate   143, 146-150, 155, 160,
                    164, 167, 169, 176, 193, 220, 236
Uranyl carbonate   178-179
U-shaped column   203-208

V

Vanadium   141, 145, 151, 165, 180, 181,
Void fraction   44, 45, 50, 51, 52, 56

W

Wet phosphoric acid (uranium from)   36, 185-187